AF564909

SIX SYSTEMS OF PHILOSOPHY

SIX SYSTEMS OF PHILOSOPHY

Editor-in-Chief

Swami Chidatman Jee Maharaj

ANMOL PUBLICATIONS PVT. LTD.

NEW DELHI - 110 002 (INDIA)

ANMOL PUBLICATIONS PVT. LTD.
H.O.: 4374/4B, Ansari Road, Daryaganj,
New Delhi-110 002 (India)
Ph.: 23278000, 23261597
B.O.: No. 1015, Ist Main Road, BSK IIIrd Stage
IIIrd Phase, IIIrd Block,
Bangalore - 560 085 (India)
Visit us at: www.anmolpublications.com

Six Systems of Philosophy

First Published, 2009
ISBN 978-81-261-3619-3

PRINTED IN INDIA

Printed at Mehra Offset Press, Delhi.

Contents

Preface

India colourful and vibrant, a land as diverse as its people. A mosaic of faiths, cultures, customs and languages that blend harmoniously to form a composite whole. One of the world's oldest living civilizations—which gave to the world—the concept of zero, the primordial sound Aum..., Yoga and Buddhism.

She is old. She is young. The two diametrically opposite, different yet unique and dynamic faces of India coexist even at this turn of the millennium. With a history of heritage, learning and civilisation dating back to almost 5000 years or perhaps even more, the modern India presents itself as an ideal country. From snow clad mountains to breezy sparkling blue sea side. From arid deserts with shifting sand dunes to lush green country sides. From ancient relics and architectural ruins dating back into antiquity to modern cities, India has it all.

To the north it is bordered by the world's highest mountain chain, where foothill valleys cover the northernmost of the country's maximum states. Further south, plateaus, tropical rain forests and sandy deserts are bordered by palm fringed beaches. Its southernmost tip is the meeting point of three Oceans.

India is a land of staggering contrast with a mingling of the tradition and modernity that is an unique experience to be in India. One will forever remember the stay in this wonderful Homeland-a land truly of mystery and enchantment. Only India can offer such an astonishing variety of contrasts.

She has given birth to numerous spiritual leaders and founders of some of the popular religions of the present day world. India probably has the most religious diversity in any country. It's the birthplace of Hinduism, Buddhism, Sikhism and Jainism and is among the few places to have a resident

Zoroastrian population. India with her veritable treasure trove of culture, mysticism, philosophy, art, music and architecture to name a few. Much of India's classical music is devotional and the North Indian Hindustani and South Indian Carnatic streams are distinct and both have a complex 'raga' framework. The legacy of dance in India is tremendous. The classical dances of India are numerous such as Kathakali of Kerala, Bharatnatyam of Tamil Nadu, Kuchipudi of Andhra Pradesh, Manipuri and Odissi from Orissa are the prominent dance forms in this country that sways to an altogether novel beat. The Yakshagana, nautanki and puppetry are ancient folk forms that live on till date. The earliest specimens of Indian painting are the ones on the walls of the Ajanta Caves dating back to 2nd century BC. The Mughals had a huge impact on Indian art. The influence of Persian art brought placid garden scenes, illustrations from myths, legends and history into Indian art. Word craft, handicrafts, architecture and sculpture all contribute to this rich and varied domain. Indian literature, both in English and in the vernacular, is ever more popular around the world.

The book aims at presenting perspectives of different aspects of Indian culture. It has topics on Indian classical music and dance, painting, ayurveda and yoga which are traditional and indigenous system of physical and mental wellbeing, language, philosophy and excellence achievements by Indian in the different branches of science and social sciences. It also includes our ancient religious and classic literature.

It is hoped that this book will be welcomed by the general readers as also scholars who are interested in the Indian heritage.

—*Editors*

1

Indian Philosophy of Six Systems

Introduction

Man's glory is not in what he is but in what he makes possible by the study of himself and nature. Philosophy is defined as the study of elements, powers, or causes and laws that explain the facts and existences. Philosophy is subjected to the influence of the race and culture and the practice of religion is nothing but a quest for the facts of inner life. In India philosophy and life are inter-linked and helped to withstand numerous external invasions and internal disturbances. Except for Charvakas who believed that materialistic gains lead to salvation, there was no place for materialistic gains in the Indian thought. Philosophy in India deals with both the cleansing of body and mind a concept well known to Indians. Indeed philosophy in India is *Aatma vidya,* knowledge of the self and "*Aatmaanam Viddhi*", know the self, sums up the Indian thought.

Although Indian philosophy (*Darshana*, to see) uses reasoning extensively yet it believes that intuition is the only way by which the ultimate truth can be known because in Indian philosophy truth is not known but realised.

Acceptance of Vedas as source of ultimate knowledge, intuition and inference unifies the various philosophical schools of India. Although all schools use words like avidya, maya, purusha and jiva, their interpretation is given differently. Another common feature of the different schools is in their non-acceptance of the Buddhist philosophy. This is probably because the various schools never tried to overturn the existing social settings where as the Buddhist preachers always spoke against the existing class (varna) system.

Periods of Indian Philosophy

Indian philosophy has always been dynamic. In Rig-Veda the religious thought of the Aryan invaders takes note of the views of the native people. In Atharva Veda vague cosmic gods are added to Gods of nature. Upanishads are realisations of things already said in the Vedas.

The Epics are the meeting points of highly philosophical thinking with the early nature worship. Since Indian philosophy has been a dynamic phenomenon, it has periods of similarities based on ideologies followed during that period. Indian philosophy can be broadly classified into

1. *The Vedic Period (1500 BC to 600 BC):* Period of Aryan invasion, Vedas and Upanishads. During this period philosophy was in its infancy with a visible conflict between superstitions and thought.
2. *The Epic Period (600 BC to 200 AD):* Period of Upanishads, epics like Ramayana and Mahabharata, Buddhism, Jainism, Shaivism and Vaishnavism.
3. *The Sutra Period (From 200 AD):As* The philosophical material available during this period was large the Sutras (or rules) were devised as shorthand scheme, since the Sutras by themselves are not intelligible, commentaries were needed. It was in this period that the six systems of Indian philosophy came into existence. It is impossible to ascertain which among the six systems is the oldest since all of them have cross-references to each other. The six Indian systems (had western counterparts)
    - *Nyaya*-Aristotle' peripatetic.
    - *Vaiseshika*-Ionian philosophyof Thales.
    - *Samkhya*-Italic philosophy of Pythogoras.
    - *Yoga*-Stoic philosophy of Zeno.

- *Purva Mimamsa*-Psychagogia of Socrates.
- *Vedanta*-Plato's philosophy

1. The Scholastic Period (upto the 16C AD): overlaps with the previous period. Great thinkers like Shankara, Kumarila Bhattar, Ramanuja were seen in this period.

The Vedic Period

Vedas are the earliest known texts of human thought. The Vedic hymns have three aspects: polytheism (Indra, Varuna etc.), monotheism (Brhaspati who represented the unification of the multitudes of gods) and monism (that there is only one supreme force of which all gods are manifestations). The notable feature of the Vedas is the conflict between primitive nature worship and an intellectual thought process trying to identify a supreme force to replace the plethora of gods. Dissatisfaction with dogmatic religion gave rise to the Upanishadic period of deep thought.

Philosophy of the Upanishads

Upanishad literally means secret teaching and form the end portion of the Vedas, called Vedanta. Since the teachings of the Upanishads are difficult to follow many schools of Vedanta arose at a later stage, each giving their own interpretation to the Upanishads. The period of the Upanishads was less scientific and more towards finding a tentative solution to satisfy the human curiosity. There are 108 Upanishads of which 10 commented by Shankaracharya are important. It is not possible to fix the dates of composition of these Upanishads that belong to pre and post Buddhist periods.

The underlying principle of this period was monism. The entire world is a manifestation of the supreme power called Brahman. Brahman is identified as "sachcitananda" (sat+chit+ananda) and is taken to be the ultimate reality. It is thought of as spiritual in nature and everything else in the world exists in and through it. *Sat* means truth that distinguishes the real Brahman from the non-being, *cit* means consciousness and indicates a spiritual nature, *ananda* means peace and refers to an all-embracing character.

Philosophy of Bhagavad Gita

Bhagavad Gita is set as a conversation between Arjuna and Sri Krishna in Mahabharata war. Although Gita is mentioned in Bhishma Parva of the Mahabharata, many believe that it is a later addition by the same author since the setting is not congenial to a long conversation but the language and style of the Mahabharata and the Gita are the same. The doctrine of Gita is essentially a theistic one.

After the Upanishadic period two schools of thought were prevalent. One was the ideal of renunciation and the other the positive approach of active work. In Gita Krishna advises Arjuna to follow the path of active work (Karma Yoga) with out anticipation of any result. He further elaborates that no work is high or low provided one does the work allotted to him based on his class (Varna). This is the concept of svadharma. Gita further elaborates that the concept of renunciation of the fruit of labour is to cleanse the heart and serve God and not to make living purposeless. Although man is not to expect the results of his deeds, he is responsible for his actions and these determine the kind of life man will live in his next life.

Influence of Some Branches of Indian Philosophy on Arts

Samkhya: We have previously seen how Samkhya classifies various qualities into Satvik, Rajasik and Tamasik. This form of classification is found in Kathakali dance form. Since this dance form depicts countless characters, each with their own qualities, classification into the three gunas becomes important. Some characters have intermediate qualities. The Satvik characters are painted green to symbolise the inner refinement and moral excellence. The gait is graceful. The Tamasik characters (like demons) are painted black and have harsh movements and make loud noises.

Yoga: Just as in yoga all Indian art forms are methods of self-realisation and not of entertainment. To achieve this controlled internalisation of senses (pratyahara) is of immense consequence. On achieving this internalisation the artist experiences the full content of life and this is the model form which he expresses through his art.

Vishistadvaita: In Indian dances the Nayika (mortal soul) yearns to be united with her Nayaka (God) and this follows the concept of Bhakti marga prescribed by Vishistadvaita.

Conclusion

Although Indian Philosophy is ancient it has always been dynamic. There has been place for theism and atheism, for the intelligentsia and the common man, for physical and mental development. The numerous schools of philosophy follow similar principles and ideas and have been responsible for the stability of the nation over millennia. As Kautilya's saying goes "Philosophy is the lamp of all sciences".

Six Systems of Indian Philosophy

The period of Gita set the tone for answers that were scientific. Strenuous attempts were made to justify by reason what faith implicitly accepts. Of the many schools of thought or Darshanas, six became famous. They are Gautama's *Nyaya*, Kaanada's *Vaiseshika*, Kapila's *Samkhy*a, Patanjali's *Yoga*, Jaimini's Purva Mimasa and Badarayana's *Uttara Mimamsa* (or Vedanta).

The six different systems of Indian philosophy have common ideas. All systems accept the existence of a real and immortal soul, distinct from the material body. They believe in the transmigration of soul from one body to the other, concept of life-death cycles. All systems except Purva Mimamsa aim at the practical achievement of salvation. They believe in jivanmukti (liberation of life from the life-death cycle) achieved by knowledge and not death.

Nyaya

The oldest existing book on Nyaya is the 'Nyaya sutra" by Gautama, is based on the theme of knowledge and logical analysis. Perception, inference, comparison and testimony are accepted as the means of knowledge.

Nyaya accepts the plurality of souls. When this inherent, permanent and unconscious soul comes in contact with sense organs it acquires intellectual, discretionary or emotional qualities. Only by deep logical analysis based on philosophical knowledge and dissolution of doubt and ignorance, man can be liberated from the life-death cycle.

Nyaya is based on:

1. Pramana-Means to acquire knowledge.
2. Prameya-Object of knowledge.
3. Samshaya-Doubt.
4. Prayojana-Purpose.
5. Drstanta-Illustrative example.
6. Siddhanta-Conclusion
7. Avayava-Constituents of a syllogism.
8. Tarka-Hypothetical argument.
9. Nirnaya-Ascertaining the truth.
10. Vada-Discussion.
11. Jalpa-Disputation.
12. Vitanda-Refutation.
13. Hetvabhasa-Fallacy in the cause.
14. Chala-guile.
15. Jati-casuistry
16. Nigrasthana-Vulnerable point of opponent's statement

Nyaya believes in the existence of a supreme God who initiates the process of world creation. Gautama defines liberation as the absolute freedom from pain. In liberation there is no feeling of bliss because soul does not have the inherent characteristics of cognition. In this state of liberation there is activity but no selfish activity in which pleasure and happiness are present but pain is absent.

Vaiseshika

Nyaya and Vaiseshika are considered sister philosophies. The earliest extant book of Vaiseshika (vishesha = particularity) is the "Vaiseshika Sutra" by Kanada. Vaiseshika proposes the atomic nature of the world *i.e.* the world consists of invisible eternal atoms (earth, water, light and air) which are incapable of further division.

The system categorises all objects of valid knowledge or padartha into six:

- Dravya (Substance)
- Guna (Quality) possess real objective experience
- Karma (Action)
- Samanya (Generality)
- Visesa (Particularity) inferred
- Samavaya (Inherence)

The commentators of the Sutras accept a seventh category called abhava (non-existence). According to Vaiseshika reality consists of substances which possess qualities and there are nine such substances namely earth, water, light, air, ether, time, space, soul and mind. Existence of soul is inferred because consciousness cannot reside in the body or sense organs but the all-pervading soul resides where the body is. The plurality of souls is inferred from their differences in status and conditions and it experiences the consequences of its deeds. Kanada does not mention God but the later commentators felt that God produced the unchangeable atoms.

Samkhya

Samkhya is considered to be the oldest among the philosophical systems dating back to about 7c BC. Kapila, the author of 'Samkhya Sutra", is considered to be the originator of this system. The "Samkhya Karika" of Ishwarakrishna is the earliest available text on Samkhya dating to about 3c AD. Samkhya's name is derived from root word Samkhya (enumeration) and is reflective than authoritative. Well-known commentaries are Gaudapada's bhasya, Vacaspati Misra's Tattwa-kaumudi, Vijnanabhiksu's Samkhya-pravacanbhasya, and Mathara's Matharavrtti.

The Samkhya system proposes the theory of evolution (prakriti-purusha) that is accepted by all other systems. The purusha (soul) of this system is unchanging and is a witness to the changes of prakriti. Hence the Samkhya system is based on dualism wherein nature (prakriti) and conscious spirit (purusha) are separate entities not derived from one another. There can be many purushas since one man can attain enlightenment while the rest do not, whereas prakriti is one. It is identified with pure objectivity, phenomenal reality, which is non-conscious.

Prakriti possess three fundamental natures; (1) The pure and fine Sattva (2) the active Rajas and (3) the coarse and heavy Tamas. Sattva accounts for thought and intelligibility, experienced psychologically as pleasure, thinking, clarity, understanding and detachment. Rajas accounts for motion, energy and activity and it is experienced psychologically as suffering, craving and attachment. Tamas accounts for restraint and inertia. It is experienced psychologically as delusion, depression and dullness.

The conscious Purusha excites the unconscious Prakriti and in this process upsets the equilibrium of the various gunas. According to Samkhya there are twenty-five tatvas which arise due to the union of purusha and prakriti. Their union is often described as the ride of a lame man with perfect sight (purusha) on the shoulders of a blind person of sure foot (prakriti). Their process of evolution is as given below and it accounts for the different tatvas. In Samkhya creation is the development of the different effects from mulaprakriti and destruction their dissolution into mulaprakriti.

Samkhya is essentially atheistic because it believes that the existence of god cannot be proved. Prakriti, the cause of evolution of world, does not evolve for itself but for Purusha- the ultimate consciousness. The self is immortal but due to ignorance (avidya) it confuses itself with the body, mind and senses. If avidya is replaced by vidya the self is free from suffering and this state of liberation is called kaivalya. Yoga is the practical side of Samkhya.

Yoga

Philosophy requires a pure body and a pure mind. Yoga is the way to achieve this. Samkhya denies the presence of god whereas in Yoga the ultimate unchanging perfect Purusha is termed god and if meditated upon he takes other purushas towards salvation. Yoga is an ancient system; in fact the excavations from Indus Valley civilisation show some yogic postures. Patanjali's *Yogasutra* is regarded as the main source of codification of yoga.

The word Yoga has a variety of meanings; it means *method* (BG), *Yoking* (RV), *conjunction of the individual and the supreme soul* (Yajnavalkya). Yoga is the methodical effort to attain perfection through the practical control of different physical and psychical elements of human nature. It gives the methods by which the body and the mind can be made to achieve vidya and hence jivanmukthi. Buddhi of Samkhya replaced by Chitt in Yoga, which undergoes modifications when in contact with the senses.

Yoga's physiology is based on a network of about 7000 Nadis (small nerves). According to Yoga the human body consists of two parts, the upper body (torso, arms and head) and the lower body (legs and feet). The centre of the human body lies at the base of the cerebro-spinal (Brahmadanda or Merudanda) column. This Merudanda has six plexus (chakras) which are the invisible to the human eye but are visible through yoga. The body is considered to be the instrument for the expression of spiritual life. In this system the physical world is not treated as unreal, instead methods to overcome the hindrances caused by the manifest world are given. There are eight such methods namely:

- Yama abstention External aids
- Niyama observance
- Aasana posture
- Pranayama breath control
- Pratyahara withdrawal of senses
- Dhaarana contemplation Internal aids
- Dhyana fixed attention
- Samadhi concentration

Yama and Niyama are the ethical preparations necessary for the practice of Yoga. To be seated comfortably is described in Aasana. (*Sthiram Sukham Aasanam*). The perfect aasana is such that the body has beauty, grace, strength and hardness. The next is breath control (pranayama) which Patanjali mentions as an optional measure since serenity of mind may be attained by cultivation of virtues or by regulation of breath.

Withdrawal of the senses and shutting of the mind to external influences (pratyahara) helps man in introspection. These five are considered to be accessories to Yoga and not themselves elements in it. Dhaarana is fixing the mind (chitt) on a particular spot. Dhyana is the resultant state of undisturbed mind. Samadhi is the condition to be passed through before attaining jivanmukthi.

Purva Mimamsa

Purva Mimamsa is earlier (purva) to Uttara (later) Mimamsa in a logical sense. By the Sutra period the Vedas were beginning to lose their glory in public opinion hence some scholars began to reanalyse the Vedas in order to defend them and justify Vedic ritualism. Mimamsa is to investigate Dharma (duty) as given in the Vedas. The Purva Mimamsa is practical than speculative. Jaimini's *Mimamsa Sutra*"s date is probably 4c BC. Sabara's commentary on *Mimamsa Sutra* is the earliest extant commentary. Kumarila, a vigorous exponent of Brahminical orthodoxy commented on *Mimamsa Sutra* and its *Bhashya*.

Jaimini accepts perception, inference and testimony but rejects intuition. He also believes that there is a connection between an act and its result since an act performed today cannot give rise to a result later, it does not give rise to some unseen result before passing away. He calls this unseen force as Apurva. Apurva is the metaphysical link between work and its result. He is unwilling to trace the origin of result to God's will since one cause cannot give multiple results.

Moksha (liberation) in Purva Mimamsa is in heaven. Purva Mimamsa gives a way to heaven but not freedom from Samsara (daily life). Since the way to liberation is defined and it is set as a goal, Purva Mimamsa goes beyond being a commentary on the Vedas to being a Darshana. It was the later philosophers of this system who brought about the change. The Purva Mimamsa was an unsatisfactory system of philosophy, which could neither explain the working of the universe nor god. Utmost importance was given to sacrifices and rituals.

Uttara Mimamsa (Vedanta Sutra)

The Vedanta philosophy is important both for its philosophical views and its close connections with the existing Hinduism. Uttara Mimamsa (also called Vedanta Sutra or Brahma Sutra) of Badarayana deals with the Brahman doctrine. It consists of 555 sutras, each with two or three words. The sutras them selves are unintelligible and leave the interpretation to the reader and hence there are varied commentaries on the Vedanta Sutra. The chief commentators are Shankara, Ramanuja, Madhava, Nimbraka and Vallabha.

The Vedanta Sutra has four chapters. The first deals with the theory of Brahman and its relation to the world. The second chapter discusses the objections raised by other systems against this view. The third chapter discusses the methods by which Brahma Vidya can be attained. The fourth chapter deals with the results of Brahma Vidya. Badarayana's view on God is that of monism. The two main commentaries of Shankara and Ramanuja are discussed below.

Advaita Vedanta of Shankara

The main exponent of advaita philosophy was Shankaracharya (8c AD). He could understand the philosophy of the Vedas by the age of eight and lived to an age of 32 by when he established four main monasteries in India, commented upon the Upanishads and the Bhagavad Gita and composed many philosophical works. Advaita literally means non-dualism and is based on the Upanishads, Brahma Sutra and Bhagavad Gita. Advaita asserts that the real self (jiva) is Brahman who is nirguna (attribute-less), nishkriya (activity-less), nir-avayava (without parts) nirupadhika (unconditional absolute) and nirvisesha (simple, homogeneos entity). The world is a manifestation of this Brahman and hence much importance is given to the Upanishadic statements like *tat tvam asi* (thou art that) and *aham brahmasmi* (I am Brahman).

Brahman is the reality behind the world and is also the Antaryami (the inner dwelling one). Liberation (moksha) is to understand that the inner self is the real Brahman. This is not a mere intellectual exercise but depends on the experiences of the person.

Although the world according to Advaita is monistic the real world perceived by man is dual. This discrepancy is attributed to the presence of Maya (illusion) or Adhyasa (superimposition) and absence of knowledge. Just as a man looking at rope mistakes it to be a snake (*rajju-sarpa bhranthi*), similarly there is no real world; it is the absence of knowledge that makes man to assume the existence of one.

The root meaning of the word *Maya* is "extraordinary and inexplicable power"; and in Shankara's Vedanta the term refers to the creative power of *Brahman* to issue forth in self-manifestation as the phenomenal world. Hence Maya and the world according to Shankara are neither real nor unreal.

Vivartavada, the philosophy that this world of variegated forms is an illusion and that the individual atma (soul) is one with God, was taught by Shankara. Here a higher reality is visible as a lower one and the cause produces the effect without undergoing any change itself. For example the same clay (Brahman) takes form of pots and toys (the variety of beings in the world) without changing its true nature.

On the practical side Advaita prescribes Yoga and Karma-sanyasa as the methods to achieve moksha. It is with the help of reasoning that avidya can be removed and liberation when alive (jivanmukthi) can be attained. Man then realises that the entire world is the manifestation of Brahman

Brahma Satya Jagan Mithya Jivo Brahmaiva Naparajh.

Vishista Advaita of Ramanuja

Ramanuja is the 11c AD founder of Vishista-advaita Vedanta (qualified monistic Vedanta) who maintained that God himself is composed of parts; individual souls and physical world comprise the body of God. Ramanuja unified a personal theist god with the single universal god of monism. This unification made the system popular among the common folk.

The sources for Ramanuja's philosophy are twofold:

- The Sanskrit Vedas, Upanishads and Puranas
- The Tamil spiritual literature which have Vedic and non-Vedic Ideas.

Ramanuja recognises three factors:

- Achit: matter which has no conscious
- Chit: Soul or those which have life and can experience
- Ishvara: God

Chit and Achit are dependent on Ishvara. Chit, Achit and Ishvara are distinct but their organic unity is accepted.

Ramanuja's Vishista Advaita consists of a seven-fold path

1. Viveka abstention
2. Vimoka freeness of mind
3. Adhyasa repetition
4. Kriya works
5. Kalyana virtuous conduct
6. Anavasada freedom from dejection
7. Anuddharasha absence of exhalation

Ramanuja's philosophy does not accept jivanmukthi. According to Ramanuja Moksha is liberation from the fetters of Samsara by seeking the rescuing love of God and the first step towards this is Nish-Kama-Karma *i.e.* practice of duty for duty's sake with out seeking for any pleasure from it and by this man has no bondage to the body. This paves the way for realisation of the soul. Karma Yoga hence becomes Jnana Yoga by following Bhakthi Yoga. Ethical religion hence gets transformed to religious philosophy.

Religion evolves as a means to narrate, to gain intellectual grasp, and to enable the seeker after truth to meditate upon the concepts so as to become free. The six systems of Hindu philosophy are concerned with intellectual analysis and sharpening of 'reason' necessary to comprehend the true nature of self, God, and universe. These six systems are the Vaisheshika, the Nyaya, the Samkhya, the Yoga, the Mimamsa, and the Vedanta. Rishis Kanada, Gotama, Kapila, Patanjali, Jaimini, and Vyasa are believed to be the earliest exponents of these systems respectively.

Although exact dates of the origin of these schools of thought are not known they are believed to have been formulated in sutras or aphorisms prior to Buddha. Many put them roughly between 600 and 200 BCE. Then the art of writing was unknown and as such these sutras were handed down from teacher to disciples by word of mouth.

To minimize the load on memory these aphorisms were rendered as short as possible, an extra dot or letter was ruthlessly deleted. However, this miserliness of words made the sutras unintelligible without commentaries and explanatory notes. Thus, we see many additions to the original text over a period of time.

There are certain common features to these six systems of thought; first and foremost is that they accept the authority of the Vedas, the feature that distinguishes them from philosophical schools of Buddhism and Jainism. This feature gives them the label of Hindu orthodox systems.

Second important feature is that, although superficially these systems seem to have contradictions amongst them, they in fact represent a progressive development from lower to higher truth. All the six schools believe in the 'Law of Karma', rebirth, and attainment of Moksha/Liberation as the highest goal of human struggle. All the systems are concerned with the nature of true Self, the realization of which through Yoga and other spiritual disciplines makes one free.

In the context of modern times, Vaisheshika is not of great importance, while Nyaya and Samkhya are studied widely for their powerful system of logic and analytical cosmology respectively. Mimamsa mostly deals with ritual portion of the Vedas, believing that sutras/verses without corresponding ritual/ Yajna is incomplete and thus do not yield desired effect.

Yoga and Vedanta have caught the attention of students of religion, scholars, as well as lay people for their practicality, rationality, and scientific basis. All Hindus now accept Vedanta as their 'living faith'.

Nyaya and Vaisheshika

These two systems are separate and independent. But for the sake of convenience and because of certain commonality these can be grouped together. Vaisheshika is older than Nyaya. Rishi Kanada or Uluka in his sutras maintains that proper object of philosophy is to focus on Dharma, virtue, so that people can prosper in life and character-Abhyudaya-as well as can attain the highest goal/good in life-liberation, Nihshreyasa.

This liberation can be attained by direct perception or knowledge of ultimate realities of Self and the universe. These ultimate realities-padarthas, or categories-are termed as 1. dravya (substance), 2. guna (qualities), 3. karma (action or motion), 4. samanya (genus), 5. vishesah (species), and 6. samavaya (relation), and additionally 7. abhava (negation).

Out of these, dravya is the basic and independent category. On this basic dravya depend all other categories.

Dravya, in turn, are nine in number. 1) The Self, 2) Manas or mind, 3. 4. 5. 6. 7.) Earth, water, air, fire, ether, 8. 9.) space and time.

Self is the substratum of consciousness. Manas when comes in contact with Self is birth, and when gets detached from it is death. Man or individual self is jivatman, and is distinct from paramatman or Supreme Self. God is the efficient cause of creation. The law of causation applies here also. Cause and effect cycle creates new world in each cycle, aarambhavada.

Nyaya deals with knowing. It has sixteen categories, dealing with the means to understand the universe. Ignorance bars the way to liberation. Ignorance results from identification of the Self with the body, the sense, and the mind. Thus we become slaves to attachment and hatred. These are the causes of our sins and sufferings. Death causes rebirth because of our ignorance of the True Self. Transcendental knowledge of our True Self is Liberation, end of cycle of birth and death, and freedom from misery.

To sum up:

Nyaya and Vaisheshika systems assert that by leading virtuous life and remaining on the side of Dharma, one will have both growth and fulfillment in life (Abhyudaya) as well as realization of the highest good-Liberation (Nishreyasa). These systems are dualistic in their conception of God (Supreme Self), Jiva (individual self), and the universe. These units have their own separate objective reality, it is said. God is seen as the Supreme Ruler of this universe, of which Jiva is a part. Existence of God is accepted as a ruler, lawgiver, controller, and governor that allows maintenance of order in the universe. According to Nyaya theory the world is more or less as we perceive it. However, the defects in the sense organs like improper or partial perception of something and the influence of fear, anticipation and other mental conditions lead to inappropriate and defective perception of the same. Nyaya regards that clear perception in general is a sound means of cognition, which discloses things to us as they really are.

As can be seen, this all is very complicated to understand and difficult write. I have written on the basis of my reading of the six systems from the book: 'The Spiritual Heritage of India' by Swami Prabhavananda.

Samkhya

Samkhya forms the philosophical basis for Yoga of Patanjali. Therefore, both Yoga and Samkhya can be grouped together, Yoga forming the practical methodology to achieve the goal. Samkhya provides rational analysis of the Truth. Its reasoning is very logical, neat, and scientific that it does not find it necessary to posit the concept of God to comprehend the Truth and freedom form suffering. Basically, Samkhya is a system based on duality of Purusha and Prakriti. While Purusha is posited as the only sentient being, ever existent, and immaterial, Prakriti is said to be the material basis of this universe, composed of three basic elements, Gunas, namely Tamas, Rajas, and Sattva.

Permutations and combinations of these gunas lead to formation of multifarious world. Not only all material objects fall under this category, but also mind and its functions of language and thoughts are composed of matter in the final analysis. Therefore, it is necessary to go beyond-transcend-the mind to reach immaterial sentient reality. This is the goal and endeavor of all human actions and activities, and Yoga Sutra of Patanjali explains and guides step by step us to this realization. First, by controlling the mind stuff-chitta-the sadhaka tries to minimize modifications of mind. And a stage is reached when single pointed mind can be transcended and the state of samadhi reached.

Yoga of Patanjali

Raja Yoga as expounded by Patanjali is the most scientific way to realize our higher consciousness. This yoga does not require belief in God, although such a belief is accepted as help in initial stage of mental concentration and control of mind. Thus, Patanjali does not deny Ishvara, as against Samkhya where necessity of God is not felt for epistemological clarity about the interrelationship between higher Self, individual self, and the universe around us.

However, the main problem with Samkhya system of thought lies with its conception of two absolutes, Purusha and Prakriti. Moreover, Samkhya maintains existence of multiple Purushas, which again puts severe limitations in accepting this philosophical analysis fully. How can two infinite ever exist! For, if Purusha and Prakriti are absolute truths in themselves, and each one is limitless and eternal, then, would not the one limit the other? And whatever is limited cannot be infinite. Despite these technicalities Samkhya system provides the most rational basis for explaining consciousness and functions of our mind. It takes clear and rationalist stand in explaining mind and thoughts as matter and therefore insentient, only Purusha being immaterial and sentient.

The whole nature exists; the whole thought process evolves only for the Purusha to gain knowledge through multiple experiences. Knowing whole of Prakriti as insentient, the Purusha is freed from the ignorance and becomes free. Purusha realizes that it was never bound; it was never unhappy. Suffering, pain and pleasure, life and death, all dualities of emotions and sense perceptions belong to Prakriti and not to Purusha. This knowledge makes one free; Purusha now shines in its own glory as absolute knowledge and consciousness, freed forever from the cycle of life and death.

Purva Mimamsa

The word Mimamsa means to analyze and understand thoroughly. The philosophical systems of karma-mimamsa and Vedanta are closely related to each other and are in some ways complimentary. Purva Mimamsa examines the teachings of the Veda in the light of karma-kanda rituals, whereas Vedanta examines the same teachings in the light of transcendental knowledge. The karma-mimamsa system is called purva-mimamsa, which means the earlier study of the Veda, and Vedanta is called uttara-mimamsa, which means the later study of the Veda (Upanishads).

Vedanta

Vedanta bases its many observations and explanations on Samkhya system. The Samkhya study and observation on cosmology are unparalleled in the realm of science and methodology. However, Vedantist does not accept multiple Purushas; Reality is one as Brahman and the rest of the universe including mind and its modifications are but superimposition upon this one Reality. The one appears as many due to the basic ignorance-avidya-related with such superimposition, which is also loosely termed as Maya. Briefly stated, the Vedanta can be put as:

Absolute Monism of Shankara

Shankara maintains that there exists but One Reality as Brahman whose nature is pure Consciousness. Naturally that One has to be eternal, all pervading and without form and attributes. If there is any second to limit the One, it cannot remain infinite. Shankara discusses the question of reality of individual soul, this world, and Ishwara from this point of view and labels them as illusory. He introduces the concept of primordial Ignorance-avidya-that deludes the human being in seeing the multifarious world when there is none! Shankara takes help of such Great Upanishadic Sayings as 'Aham Brahmasmi (I am Brahman)' and 'Tat Tvam Asi' (Thou Art

That). Final Liberation comes when this knowledge of unity of individual soul and eternal Soul is established through meditation and samadhi, transcendental knowledge. Shankara is strong proponent of Jnana Yoga.

Vishishtha Advaita or Qualified Monism of Ramanuja

Vishishtha Advaita or qualified monism of Ramanuja is a philosophy of religion; and therefore it gives a synthetic view of the spiritual experiences of God or Brahman. It affirms the Upanishadic truth by realizing Brahman everything else is realized. The main theme is that it is Brahman Itself who has become this whole universe. Thus Ramanuja accepts the reality of world as the manifestation of Brahman, differing here with Maya theory of Shankara. Vedanta accepts rational approach to Truth but goes beyond the reason and rationality without contradicting them. Heavily depending upon theory of Karma, this philosophy applies the law of cause and effect to moral experiences. It brings to light the inner working of righteousness of God and affirms the impossibility of cruelty and bias in Divine nature. As you sow, so would you reap! This law applies to explain suffering for many and comfort for others. But the Grace of God, the most benevolent being, transfigures the rigorous law of karma and thus grace or kripa becomes the ruling principle of religion. Thus Ramanuja added concepts of devotion, worship, and faith as new dimensions to Vedanta. Ramanuja does not accept the impersonal Brahman without attributes of Shankara, but rather an eternal personal Brahman, the repository of all blessed qualities.

Dvaita of Madhva

Dvaita of Madhvacharya propounds the basic theory that Jiva (individual soul), Jagat (manifest world), and Jagadish (Ruler of this universe-God) all the three are real, eternal and separate. One cannot become the other. Devotion and faith in the goodness of the Lord is the only means to reach near Him. For them God may be Hari or Vishnu with form and divine attributes. Madhvacharya found that real cause of bondage is superimposition of 'doership' (I have done this, I do this, I shall do that!) upon individual self. But the fact is that Jiva is dependent on God for all his actions. He saw that the only way of appreciating the highest Truth is recognition of this reality and actuality of this world in a healthy sense. Jiva through devotion, worship, and surrender to Almighty should understand the grace and power of the God, and that would bring peace and tranquillity to individual self and the world order.

Dvaita is totally opposed to Maya theory of Shankara that denies the very existence of Jiva and the world as illusory. Thus, Madhva identifies the Brahman of the Upanishads with Vishnu, and forcefully argues against the dichotomy of Shrutis as claimed by Sri Shankaracharya, saying that such arbitration of apaurusheya scripture is unacceptable both logically and spiritually. He also emphasizes that it is important to understand and specifically reject other schools' precepts, and hence devotes much time to nitpicking analyses and denunciations of other doctrines.

Conclusion

All Indian philosophy considers ignorance as a barrier to liberation. This ignorance results from false identification of the Self with body-mind-sense complex. Thus we are entangled in the mesh of attachment and hatred that invariably leads to all our selfish and therefore sinful acts and sufferings. Only when one attains true knowledge of the true Self (God, Brahman, Consciousness) one is freed from the slavery of the senses and all sufferings come to an end.

Throughout its long and largely unrecorded history, Indian thought preserved its central concern with ontology and epistemology, with noetic psychology as the indispensable bridge between metaphysics and ethics, employing introspection and self-testing as well as logical tools, continually confronting the instruments of cognition with the fruits of contemplation. Through its immemorial oral teachings and a vast variety of written texts, the fusion of *theoria* and *praxis*, theory and practice, was never sacrificed to the demands of academic specialization or the compartmentalization of human endeavour. Diverse schools of thought shared the conviction that true understanding must flow from the repeated application of received truths. Coming to know is a dynamic, dialectical process in which thought stimulates contemplation and regulates conduct, and in turn is refined by them. Although an individual who would be healthy and whole thinks, feels and acts, gnosis necessarily involves the fusion of thought, will and feeling, resulting in *metanoia*, a radically altered state of being. The Pythagorean conception of a philosopher as a lover of wisdom is close to the standpoint of an earnest seeker of truth in the Indian tradition.

Indian thought did not suffer the traumatic cognitive disruption caused by the emergence of ecclesiastical Christianity in the Mediterranean world, where an excessive concern with specification of rigid belief, sanctioned and safeguarded by an institutional conception of religious authority and censorship, sundered thought and action to such an extent that it became common to think one way and act in another with seeming impunity. The chasms which opened up between thought, will and feeling provided fertile soil for every kind of psychopathology, in part because such a fragmentation of the human being engenders inversions, obsessions and even perversities, and also in part because for a thousand years it has been virtually impossible to hold up a credible paradigm of the whole and healthy human being. The philosophical quest became obscured in the modern West by the linear succession of schools, each resulting from a violent reaction to its predecessors, each claiming to possess the Truth more or less exclusively, and often insisting upon the sole validity of its method of proceeding. The slavish concern with academic respectability and the fear of anathemization resulted in the increasing alienation of thought from being, of cognition from conduct, and philosophical disputation from the problems of daily life.

Indian thought did not spurn the accumulated wisdom of its ancients in favour of current fashions and did not experience a violent disruption of its traditional hospitality to multiple standpoints. The so-called *astika* or orthodox schools found no difficulty in combining their veneration of the Vedic hymns with a wide and diverse range of views, and even the *nastika* or heterodox schools, which repudiated the canonical "authority" of the Vedas, retained much of Vedic and Upanishadic metaphysics and almost the whole of their psychology and ethics. Indian philosophical schools could not see themselves as exclusive bearers of the total Truth. They emerged together from a long-standing and continuous effort to enhance our common understanding of God, Man and Nature, and they came to be considered as *darshanas* or paradigmatic standpoints shedding light from different angles on noumenal and phenomenal realities. They refrained from claiming that any illumination which can be rendered in words–or even in thoughts–can be either final or complete.

The Six Schools

"It may be pointed out here that a system of philosophy however lofty and true it may be should not be expected to give us an absolutely correct picture of the transcendent truths as they really exist. Because philosophy works through the medium of the intellect and the intellect has its inherent limitations, it cannot understand or formulate truths which are beyond its scope.... We have to accept these limitations when we use the intellect as an instrument for understanding and discovering these truths in the initial stages. It is no use throwing away this instrument, poor and imperfect though it is, because it gives us at least some help in organizing our effort to know the truth in the only way it can be known–by Self-realization." (I. K. Taimni)

The ageless and dateless Vedas, especially the exalted hymns of the Rig Veda, have long been esteemed as the direct expression of what gods and divine seers, rishis or immortal sages, saw when they peered into the imperishable center of Being which is also the origin of the entire cosmos. The Upanishads (from *upa*, *ni* and *sad*, meaning "to sit down near" a sage or guru), included in the Vedas, constitute the highest transmission of the fruits of illumination attained by these rishis. Often cast in the form of memorable dialogues between spiritual teachers and disciples, they represent rich glimpses of truth, not pieced together from disparate intellectual insights, but as they are at once revealed to the divine eye, *divya chakshu*, which looks into the core of Reality, freely intimated in idioms, metaphors and mantras suited to the awakening consciousness and spiritual potentials of diverse disciples. However divergent their modes of expression, they are all addressed to those who are ready to learn, willing to meditate deeply, and seek greater self-knowledge through intensive self-questioning. The Upanishads do not purport to provide discursive knowledge, conceptual clarification or speculative dogmas, but rather focus on the fundamental themes which concern the soul as a calm spectator of the temporal succession of states of mind from birth to death, seeking for what is essential amidst the ephemeral, the enduring within the transient, the abiding universals behind the flux of fleeting appearances.

From this standpoint, they are truly therapeutic in that they heal the sickness of the soul caused by passivity, ignorance and delusion. This ignorance is not that of the malformed or malfunctioning personality, maimed by childhood traumas or habitual vices. It is the more fundamental ignorance (avidya) of the adroit and well-adapted person who has learnt to cope with the demands of living and fulfil his duties in the world at a certain level without however, coming to terms with the causes of his longings and limitations, his dreams and discontinuities, his entrenched expectations and his hidden potentials.

The sages spoke to those who had a measure of integrity and honesty and were willing to examine their presuppositions, but lacked the fuller vision and deeper wisdom that require a sustained search and systematic meditation. For such an undertaking, mental clarity, moral sensitivity, relaxed self-control and spiritual courage are needed, as well as a willingness to withdraw for a period from worldly concerns. The therapeutics of self-transcendence is rooted in a recondite psychology which accommodates the vast spectrum of self-consciousness, different levels of cognition and degrees of development, reaching up to the highest conceivable self-enlightenment.

Upanishadic thought presupposed the concrete and not merely conceptual continuity of God, Nature and Man. Furthermore, Man is the self-conscious microcosm of the macrocosm, where the part is not only inseparably one with the whole but also reflects and resonates with it. Man could neither be contemplated properly nor fully comprehended in any context less than the entirety of visible and invisible Nature, and so too, ethics, logic and psychology could not be sundered from metaphysics. "Is," the way things are, is vitally linked to "must," the ways things must be, as well as to "ought," the way human beings should think and act, through "can," the active exploration of human potentialities and possibilities, which are not different, save in scope and degree, from cosmic potencies. A truly noetic psychology bridges metaphysics and ethics through a conscious mirroring of *rita*, ordered cosmic harmony, in dharma, righteous human conduct that freely acknowledges what is due to each and every aspect of Nature, including all humanity, past, present and future.

The ancient sages resolved the One-many problem at the mystical, psychological, ethical and social levels by affirming the radical metaphysical and spiritual unity of all life, whilst fully recognizing (and refusing to diminish through any form of reductionism) the immense diversity of human types and the progressive awakenings of human consciousness at different stages of material evolution and spiritual involution. The immemorial pilgrimage of humanity can be both universally celebrated and act as a constant stimulus to individual growth. Truth, like the sun shining over the summits of a Himalayan range, is one, and the pathways to it are as many and varied as there are people to tread them.

As if emulating the sculptor's six perspectives to render accurately any specific form in space, ancient Indian thinkers stressed six darshanas, which are sometimes called the six schools of philosophy. These are astika or orthodox in that they all find inspiration in different ways in the Vedas. And like the sculptor's triple set of perspectives–front-back, left side-right side, top-bottom–the six darshanas have been seen as three complementarities, polarized directions that together mark the trajectory of laser light through the unfathomable reaches of ineffable wisdom. Each standpoint has its integrity and coherence in that it demands nothing less than the deliberate and radical reconstitution of consciousness from its unregenerate and unthinking modes of passive acceptance of the world. Yet none can claim absoluteness, finality or infallibility, for such asseverations would imply that limited conceptions and discursive thought can capture ultimate Reality. Rather, each darshana points with unerring accuracy towards that cognition which can be gained only by complete assimilation, practical self-transformation and absorption into it. At the least, every darshana corresponds with a familiar state of mind of the seeker, a legitimate and verifiable mode of cognition which makes sense of the world and the self at some level.

All genuine seekers are free to adopt any one or more of the darshanas at any time and even to defend their chosen standpoint against the others but they must concede the possibility of synthesizing and transcending the six standpoints in a seventh mode which culminates in taraka, transcendental, self-luminous gnosis, the goal of complete enlightenment often associated with the secret, incommunicable way of *buddhiyoga* intimated in the fourth, seventh and eighteenth chapters of the Bhagavad Gita.

Although scholars have speculated on the sequential emergence of the darshanas, and though patterns of interplay can be discerned in their full flowering, their roots lie in the ancient texts and they arise together as distinctive standpoints. It has also been held that the six schools grew out of sixty-two systems of thought lost in the mists of antiquity. At any rate, it is generally agreed that each of the later six schools was inspired by a sage and teacher who struck the keynote which has reverberated throughout its growths refinement and elaboration. As the six schools are complementary to each other, they are traditionally viewed as the six branches of a single tree. All six provide a theoretical explanation of ultimate Reality and a practical means of emancipation. The oldest are Yoga and Sankhya, the next being Vaishesika and Nyaya, and the last pair are Purva Mimansa and Vedanta (sometimes called Uttara Mimansa). The founders of these schools are considered to be Patanjali of Yoga, Kapila of Sankhya, Kanada of Vaishesika, Gautama of Nyaya, Jaimini of Purva Mimansa and Vyasa of Vedanta, though the last is also assigned to Badarayana. All of them propounded the tenets of their philosophical systems or schools in the form of short sutras, whose elucidation required and stimulated elaborate commentaries. Since about 200 C.E., a vast crop of secondary works has emerged which has generated some significant discussions as well as a welter of scholastic disputation and didactic controversies, moving far away from praxis into the forests of theoria, or reducing praxis to rigid codes and theoria to sterile formulas.

At the same time, there has remained a remarkable vitality to most of these schools, owing to their transmission by long lineages which have included many extraordinary teachers and exemplars. This cannot be recovered merely through the study of texts, however systematic and rigorous, in a philosophical tradition which is essentially oral, even though exceptional powers of accurate recall have been displayed in regard to the texts.

Nyaya and Vaishesika

Nyaya and Vaishesika are schools primarily concerned with analytic approaches to the objects of knowledge, using carefully tested principles of logic. The word *nyaya* suggests that by which the mind reaches a conclusion, and since the word also means "right" or "just," Nyaya is the science of correct thinking. The founder of this school, Gautama, lived about 150 B.C.E., and its source-book is the *Nyaya Sutra*. Whilst knowledge requires an object, a knowing subject and a state of knowing, the validity of cognition depends upon *pramana*, the means of cognition. There are four acceptable *pramanas*, of which *pratyaksha*–direct perception or intuition–is most important. Perception requires the mind, manas, to mediate between the self and the senses, and perception may be determinate or indeterminate. Determinate perception reveals the class to which an object of knowledge belongs, its specific qualities and the union of the two. Indeterminate perception is simple apprehension without regard to genus or qualities. In the Nyaya school, indeterminate perception is not knowledge but rather its prerequisite and starting-point.

Anumana or inference is the second pramana or means of cognition. It involves a fivefold syllogism which includes a universal statement, an illustrative example and an application to the instance at hand. *Upamana* is the apt use of analogy, in which the similarities which make the analogy come alive are essential and not superficial. Shabda, sound or verbal expression, is the credible testimony of authority, which requires not uncritical acceptance but the thoughtful consideration of words, meanings and the modes of reference. As the analytic structure of Nyaya logic suggests, its basic approach to reality is atomistic, and so the test of claims of truth is often effectiveness in application, especially in the realm of action. Typically, logical discussion of a proposition takes the form of a syllogism with five parts: the proposition (pratijna) the cause (hetu), the exemplification (drishtanta), the recapitulation (upanaya) and the conclusion (nigamana).

However divergent their views on metaphysics and ethics, all schools accept and use Nyaya canons of sound reasoning. A thorough training in logic is required not only in all philosophical reasoning, exposition and disputation, but it is also needed by those who seek to stress mastery of praxis over a lifetime and thereby become spiritual exemplars. This at once conveys the enormous strength of an immemorial tradition as well as the pitiable deficiencies of most professors and pundits, let alone the self-styled so-called exoteric gurus of the contemporary East. Neither thaumaturgic wonders nor mass hypnosis can compensate for mental muddles and shallow thinking; indeed, they become insuperable obstacles to even a good measure of gnosis and noetic theurgy, let alone authentic enlightenment and self-mastery.

The Vaishesika school complements Nyaya in its distinct pluralism. Its founder, Kanada, also known as Kanabhaksha, lived around 200 C.E., and its chief work is the *Vaishesika Sutra*. Its emphasis on particulars is reflected in its name, since *vishesha* means "particularity," and it is concerned with properly delineating the categories of objects of experience. These objects of experience, padarthas, are six: substance (dravya), quality (guna), and karma or movement and activity (forming the triplicity of objective existence), and generality (samanya), particularity (vishesha) and samavayi or inherence (forming a triad of modes of intellectual discernment which require valid logical inference). A seventh object of experience, non-existence (shunya), was eventually added to the six as a strictly logical necessity. The Vaishesika point of view recognizes nine irreducible substances: earth, water, air, fire, aether (akasha), time, space, self and mind, all of which are distinct from the qualities which inhere in them. The self is necessarily a substance–a substrate of qualities–because consciousness cannot be a property of the physical body, the sense-organs or the brain-mind. Although the self as a substance must be everywhere pervasive, its everyday capacity for feeling, willing and knowing is focussed in the bodily organism.

Since the self experiences the consequences of its own deeds, there is, according to Vaishesika, a plurality of souls, each of which has its vishesha, individuality or particularity. What we experience is made up of parts, and is non-eternal, but the ultimate components–atoms–are eternal. Individuality is formed by imperceptible souls and certain atoms, which engender the organ of thought. At certain times, during immense cosmogonic cycles, nothing is visible, as both souls and atoms are asleep, but when a new cycle of creation begins, these souls reunite with certain atoms. Gautama asserted that even during incarnated existence, emancipation may be attained through ascetic detachment and the highest stages of contemplative absorption or samadhi.

Though the Vaishesika school wedded an atomistic standpoint to a strict atheism, over time thinkers accepted a rationalistic concept of Deity as a prime mover in the universe, a philosophical requisite acceptable to Nyaya. The two schools or systems were combined by Kusumanjali of Udayana about 900 C.E. in his proof of the existence of God. Since then, both schools have been theistic. The Jains claim early parentage for the Vaishesika system, and this merely illustrates what is very common in the Indian tradition, that innovators like Gautama and Kanada were reformulating an already ancient school rather than starting *de novo*.

Purva Mimansa

The Purva Mimansa of Jaimini took as its point of departure neither knowledge nor the objects of experience, but dharma, duty, as enjoined in the Vedas and Upanishads. As the accredited sources of dharma, these sacred texts are not the promulgations of some deity who condescended to step into time and set down principles of correct conduct. Rather, the wisdom in such texts is eternal and uncreate, and true rishis have always been able to see them and to translate that clear vision into mantric sounds and memorable utterances.

Hence Mimansa consecrates the mind to penetrating the words which constitute this sacred transmission. Central to the Mimansa school is the theory of self-evidence–svata pramana: truth is its own guarantee and the consecrated practice of faith provides its own validation. Repeated testings will yield correct results by exposing discrepancies and validating real cognitions. There is a recognizable consensus amidst the independent visions of great seers, and each individual must recognize or rediscover this consensus by proper use and concentrated enactment of mantras and hymns. Every sound in the fifty-two letters of Sanskrit has a cosmogonic significance and a theurgic effect. Inspired mantras are exact mathematical combinations of sounds which emanate potent vibrations that can transform the magnetic sphere around the individual as well as the magnetosphere of the earth. Self-testing without self-deception can become a sacred activity, which is *sui generis*.

From the Mimansa perspective, every act is necessarily connected to perceptible results. One might say that the effects are inherent in the act, just as the fruit of the tree is in the seed which grew and blossomed. There is no ontological difference between act and result, for the apparent gap between them is merely the consequence of the operation of time. Since the fruit of a deed may not follow immediately upon the act, or even manifest in the same lifetime, the necessary connection between act and result takes the form of apurva, an unseen force which is the unbreakable link between them. This testable postulate gives significance to the concept of dharma in all its meanings–"duty," "path," "teaching," "religion," "natural law," "righteousness," "accordance with cosmic harmony"–but it cannot by itself secure complete liberation from conditioned existence. Social duties are important, but spiritual duties are even more crucial, and the saying "To thine own self be true" has an array of meanings reaching up to the highest demands of soul-tendance.

In the continual effort to work off past karma and generate good karma, there is unavoidable tension between different duties, social and spiritual. The best actions, paradigmatically illustrated in Vedic invocations and rituals, lead to exalted conditions, even to some heavenly condition or blissful state. Nonetheless, as the various darshanas interacted and exchanged insights, Mimansa came to consider the highest action as resulting in a cessation of advances and retreats on the field of merit, whereby dharma and adharma were swallowed up in a sublime and transcendental state of unbroken awareness of the divine.

In striving to penetrate the deepest arcane meaning of the sacred texts, Mimansa thinkers accepted the four pramanas or modes of knowledge set forth in Nyaya, and added two others: arthapatti or postulation, and abhava or negation and non-existence. They did this in part because, given their view of the unqualified eternality of the Vedas, they held that all cognition is valid at some level and to some degree. There can be no false knowledge; whatever is known is necessarily true. As a consequence, they saw no reason to prove the truth of any cognition. Rather, they sought to demonstrate its falsity, for if disproof were successful, it would show that there had been no cognition at all.

The promise of gnosis rests upon the sovereign method of falsifiability rather than a vain attempt to seek total verification in a public sense. Shifting the onus of proof in this way can accommodate the uncreate Vedas, which are indubitably true and which constitute the gold standard against which all other claims to truth are measured. Mimansa rests upon the presupposition of the supremacy of Divine Wisdom, the sovereignty of the Revealed Word and the possibility of its repeated realization. Even among those who cannot accept the liturgical or revelatory validity and adequacy of the Vedas, the logic of disproof can find powerful and even rigorous application. As a method, it became important to the philosophers of Vedanta.

Vedanta (Uttara Mimansa)

Vedanta, meaning "the end or goal of the Vedas," sometimes also called Uttara Mimansa, addresses the spiritual and philosophical themes of the Upanishads, which are considered to complete and form the essence of the Vedas. Badarayana's magisterial *Brahma Sutras* ordered the Upanishadic Teachings in a logically coherent sequence which considers the nature of the supreme brahman, the ultimate Reality, and the question of the embodiment of the unconditioned Self. Each of the five hundred and fifty-five sutras (literally, "threads") are extremely short and aphoristic, requiring a copious commentary to be understood.

In explaining their meaning, various commentators presented Vedantic doctrines in different ways. Shankaracharya, the chief of the commentators and perhaps the greatest philosopher in the Indian tradition, espoused the advaita, non-dual, form of Vedanta, the purest form of monism, which has never been excelled. He asked whether in human experience there is anything which is impervious to doubt. Noting that every object of cognition–whether dependent on the senses, the memory or pure conceptualization–can be doubted, he recognized in the doubter that which is beyond doubt of any kind. Even if one reduces all claims to mere avowals–bare assertions about what one seems to experience–there nonetheless remains that which avows. It is proof of itself, because nothing can disprove it. In this, it is also different from everything else, and this difference is indicated by the distinction between subject and object. The experiencing Self is subject; what it experiences is an object. Unlike objects, nothing can affect it: it is immutable and immortal.

For Shankara, this Self (atman) is sat-chit-ananda, being or existence, consciousness or cognition, and unqualified bliss. If there were no world, there would be no objects of experience, and so although the world as it is experienced is not ultimately real, it is neither abhava, non-existent, nor shunya, void. Ignorance is the result of confusing atman, the unconditioned subject, with anatman, the external world. From the standpoint of the cosmos, the world is subject to space, time and causality, but since these categories arise from nascent experience, they are inherently inadequate save to point beyond themselves to the absolute, immutable, self-identical brahman, which is absolute Being (sat). Atman is brahman, for the immutable singularity of the absolute subject, the Self, is not merely isomorphic, but radically identical with the transcendent singularity of the ultimate Reality.

Individuals who have yet to realize this fundamental truth, which is in fact the whole Truth, impose out of ignorance various attitudes and conceptions on the world, like the man who mistakes an old piece of rope discarded on the trail for a poisonous serpent. He reacts to the serpent, but his responses are inappropriate and cause him to suffer unnecessarily, because there is no serpent on the trail to threaten him. Nonetheless, the rope is there. For Shankara, the noumenal world is real, and when a person realizes its true nature, gaining wisdom thereby, his responses will be appropriate and cease to cause suffering. He will realize that he is the atman and that the atman is brahman.

Although brahman is ultimately nirguna, without qualities, the aspirant to supreme knowledge begins by recognizing that the highest expression of brahman to the finite mind is Ishvara, which is saguna brahman, Supreme Reality conceived through the modes of pure logic. Taking Ishvara, which points beyond itself to That (Tat), as his goal and paradigm, the individual assimilates himself to Ishvara through the triple path of ethics, knowledge and devotion–the karma, jnana and bhakti yogas of the Bhagavad Gita–until moksha, emancipation and self-realization, is attained. For Shankara, moksha is not the disappearance of the world but the dissolution of avidya, ignorance.

Ramanuja, who lived much later than Shankara, adopted a qualified non-dualism, Vishishtadvaita Vedanta, by holding that the supreme brahman manifests as selves and matter. For him, both are dependent on brahman, and so selves, not being identical with the Ultimate, always retain their separate identity. As a consequence, they are dependent on brahman, and that dependency expresses itself self-consciously as bhakti or devotion. In this context, however, the dependence which is manifest as bhakti is absurd unless brahman is thought to be personal in some degree, and so brahman cannot be undifferentiated. Emancipation or freedom is not union with the divine, but rather the irreversible and unwavering intuition of Deity.

The Self is not identical with brahman, but its true nature is this intuition, which is freedom. Faith that brahman exists is sufficient and individual souls are parts of brahman, who is the creator of universes. Yet brahman does not create anything new; what so appears is merely a modification of the subtle and the invisible to the gross which we can see and sense. Because we can commune with this God by prayer, devotion and faith, there is the possibility of human redemption from ignorance and delusion. The individual is not effaced when he is redeemed; he maintains his self-identity and enjoys the fruits of his faith.

About a century and a half after Ramanuja, Madhava promulgated a dualistic (dvaita) Vedanta, in which he taught that brahman, selves and the world are separate and eternal, even though the latter two depend forever upon the first. From this standpoint, brahman directs the world, since all else is dependent, and is therefore both transcendent and immanent. As that which can free the self, brahman is identified with Vishnu. Whereas the ultimate Reality or brahman is neither independent (svatantra) nor dependent (paratantra), God or Vishnu is independent, whereas souls and matter are dependent. God did not cause the cosmos but is part of it, and by his presence keeps it in motion.

Individual souls are dependent on brahman but are also active agents with responsibilities which require the recognition of the omnipresence and omnipotence of God. For the individual self, there exists either the bondage which results from ignorance and the karma produced through acting ignorantly, or release effected through the adoration, worship and service of Deity.

The self is free when its devotion is pure and perpetual. Although the later forms of Vedanta lower the sights of human potentiality from the lofty goal of universal self-consciousness and conscious immortality taught by Shankaracharya, they all recognize the essential difference between bondage and freedom.

The one is productive of suffering and the other offers emancipation from it. But whereas for Shankara the means of emancipation is wisdom (jnana) as the basis of devotion (bhakti) and nishkama karma or disinterested action, the separation between atman and brahman is crucial for Ramanuja and necessitates total bhakti, whilst for Madhava there are five distinctions within his dualism–between God and soul, God and matter, soul and matter, one form of matter and another, and especially between one soul and another–thus requiring from all souls total obeisance to the omnipresent and omnipotent God.

Suffering is the starting point of the Sankhya darshana which provides the general conceptual framework of Yoga philosophy. Patanjali set out the Taraka Raja Yoga system, linking transcendental and self-luminous wisdom (taraka) with the alchemy of mental transformation, and like the exponents of other schools, he borrowed those concepts and insights which could best delineate his perspective. Since he found Sankhya metaphysics useful to understanding, like a sturdy boat used to cross a stream and then left behind when the opposite bank has been reached, many thinkers have traditionally presented Samkhya as the theory for which Yoga is the practice. This approach can aid understanding, providing one recognizes from the first and at all times that yoga is the path to metaconsciousness, for which no system of concepts and discursive reasoning, however erudite, rigorous and philosophical, is adequate. More than any other school or system, Yoga is essentially experiential, in the broadest, fullest and deepest meaning of that term.

Samkhya

The term "Samkhya" is ultimately derived from the Sanskrit root khya, meaning "to know," and the prefix san, "exact."

Exact knowing is most adequately represented by Samkhya, "number," and since the precision of numbers requires meticulous discernment, Samkhya is that darshana which involves a thorough discernment of reality and is expressed through the enumeration of diverse categories of existence. Philosophically, Samkhya is dualistic in its discernment of the Self (purusha) from the non-self (prakriti).

In distinguishing sharply between purusha, Self or Spirit, on the one hand, and prakriti, non-self or matter, on the other, the Samkhya standpoint requires a rigorous redefinition of numerous terms used by various schools. Even though later Samkhya freely drew from the Vedic-Upanishadic storehouse of wisdom which intimates a rich variety of philosophical views, its earliest concern does not appear to have been philosophical in the sense of delineating a comprehensive conceptual scheme which describes and explains reality. Early Samkhya asked, "What is real?" and only later on added the question, "How does it all fit together?"

Enumerations of the categories of reality varied with individual thinkers and historical periods, but the standard classification of twenty-five tattvas or fundamental principles of reality is useful for a general understanding of the darshana. Simply stated, Samkhya holds that two radically distinct realities exist: purusha, which can be translated "Spirit," "Self" or "pure consciousness," and mulaprakriti, or "pre-cosmic matter," "non-self" or "materiality." Nothing can be predicated of purusha except as a corrective negation; no positive attribute, process or intention can be affirmed of it, though it is behind all the activity of the world.

It might be called the Perceiver or the Witness, but, strictly speaking, no intentionality can be implied by these words, and so purusha cannot be conceived primarily as a knower. Mulaprakriti, however, can be understood as pure potential because it undergoes ceaseless transformation at several levels. Thus, of the twenty-five traditional tattvas, only these two are distinct. The remaining twenty-three are transformations or modifications of mulaprakriti. Purusha and mulaprakriti stand outside conceptual cognition, which arises within the flux of the other tattvas. They abide outside space and time, are simple, independent and inherently unchanging, and they have no relation to one another apart from their universal, simultaneous and mutual presence.

Mulaprakriti is characterized by three qualities or gunas: sattva or intelligent and noetic activity, rajas or passionate and compulsive activity, and tamas or ignorant and impotent lethargy, represented in the Upanishads by the colours white, red and black. If mulaprakriti were the only ultimate reality, its qualities would have forever remained in a homogeneous balance, without undergoing change, evolution or transformation. Since purusha is co-present with mulaprakriti, the symmetrical homogeneity of mulaprakriti was disturbed, and this broken symmetry resulted in a progressive differentiation which became the world of ordinary experience. True knowledge or pure cognition demands a return to that primordial stillness which marks the utter disentanglement of Self from non-self.

The process which moved the gunas out of their perfect mutual balance cannot be described or even alluded to through analogies, in part because the process occurred outside space and time (and gave rise to them), and in part because no description of what initiated this universal transformation can be given in the language of logically subsequent and therefore necessarily less universal change. In other words, all transformation known to the intellect occurs in some context–minimally that of the intellect itself–whilst the primordial process of transformation occurred out of all context, save for the mere co-presence of purusha and mulaprakriti.

This imbalance gave rise, first of all, logically speaking, to mahat or buddhi. These terms refer to universal consciousness, primordial consciousness or intellect in the classical and neo-Platonic sense of the word. Mahat in turn gave rise to ahankara, the sense of "I" or egoity. (Ahankara literally means "I-making.") Egoity as a principle or tattva generated a host of offspring or evolutes, the first of which was manas or mind, which is both the capacity for sensation and the mental ability to act, or intellectual volition. It also produced the five buddhindriyas or capacities for sensation: shrota (hearing), tvac (touching), chaksus (seeing), rasana (tasting) and ghrana (smelling). In addition to sensation, ahankara gave rise to their dynamic and material correlates, the five karmendriyas or capacities for action, and the five tanmatras or subtle elements. The five karmendriyas are vach (speaking), pani (grasping), pada (moving), payu (eliminating) and upastha (procreating), whilst the five tanmatras include shabda (sound), sparsha (touch), rupa (form), rasa (taste) and gandha (smell). The tanmatras are called "subtle" because they produce the mahabhutas or gross elements which can be perceived by ordinary human beings. They are akasha (aether or empirical space), vayu (air), tejas (fire, and by extension, light), ap (water) and prithivi (earth).

This seemingly elaborate system of the elements of existence (tattvas) is a rigorous attempt to reduce the kaleidoscope of reality to its simplest comprehensible components, without either engaging in a reductionism which explains away or denies what does not fit its classification, or falling prey to a facile monism which avoids a serious examination of visible and invisible Nature. Throughout the long history of Samkhya thought, enumerations have varied, but this general classification has held firm. Whilst some philosophers have suggested alternative orders of evolution, for instance, making the subtle elements give rise to the capacities for sensation and action, Ishvarakrishna expressed the classical consensus in offering this classification of twenty-five tattvas.

Once the fundamental enumeration was understood, Samkhya thinkers arranged the tattvas by sets to grasp more clearly their relationships to one another. At the most general level, purusha is neither generated nor generating, whilst mulaprakriti is ungenerated but generating. Buddhi, ahankara and the tanmatras are both generated and generating, and manas, the buddhindriyas, karmendriyas and mahabhutas are generated and do not generate anything in turn. In terms of their mutual relationships, one can speak of kinds of tattvas and indicate an order of dependence from the standpoint of the material world.

No matter how subtle and elaborate the analysis, however, one has at best described ways in which consciousness functions in prakriti, the material world. If one affirms that purusha and prakriti are radically and fundamentally separate, one cannot avoid the challenge which vexed Descartes: how can res cogitans, thinking substance, be in any way connected with res extensa, extended (material) substance? Samkhya avoided the most fundamental problem of Cartesian dualism by willingly admitting that there can be no connection, linkage or interaction between purusha and prakriti.

Since consciousness is a fact, this exceptional claim involved a redefinition of consciousness itself. Consciousness is necessarily transcendent, unconnected with prakriti, and therefore it can have neither cognitive nor intuitive awareness, since those are activities which involve some center or egoity and surrounding field from which it separates itself or with which it identifies. Egoity or perspective requires some mode of action, and all action involves the gunas, which belong exclusively to prakriti. Consciousness, purusha, is mere presence, sakshitva, without action, dynamics or content. Awareness, chittavritti, is therefore a function of prakriti, even though it would not have come into being–any more than anything would have evolved or the gunas would have become unstable–without the universal presence of purusha. Thus it is said that purusha is unique in that it is neither generated nor generating, whereas all other tattvas are either generating, generated or both.

In this view, mind is material. Given its capacity for awareness, it can intuit the presence of purusha, but it is not that purusha. All mental functions are part of the complex activity of prakriti. Consciousness is bare subjectivity without a shadow of objective content, and it cannot be said to have goals, desires or intentions. Purusha can be said to exist (sat)–indeed, it necessarily exists–and its essential and sole specifiable nature is chit, consciousness. Unlike the Vedantin atman, however, it cannot also be said to be ananda, bliss, for purusha is the pure witness, sakshi, with no causal connection to or participation in prakriti. Yet it is necessary, for the gunas could not be said to be active save in the presence of some principle of sentience. Without purusha there could be no prakriti. This is not the simple idealistic and phenomenological standpoint summarized in Berkeley's famous dictum, esse est percipi, "to be is to be perceived." Rather, it is closer to the recognition grounded in Newtonian mechanics that, should the universe achieve a condition of total entropy, it could not be said to exist, for there would be no possibility of differentiation in it. Nor could its existence be denied. The presence of purusha, according to Samkhya, is as necessary as is its utter lack of content.

Given the distinction between unqualified, unmodified subjectivity as true or pure consciousness, and awareness, which is the qualified appearance of consciousness in the world, consciousness appears as what it cannot be. It appears to cause and initiate, but cannot do so, since purusha cannot be said to be active in any sense; it appears to entertain ideas and chains of thought, but it can in reality do neither. Rather, the action of the gunas appears as the activity of consciousness until the actual nature of consciousness is realized. The extreme break with previous understanding resulting from this realization–that consciousness has no content and that content is not conscious–is emancipation, the freeing of purusha from false bondage to prakriti. It is akin to the Vedantin realization of atman free of any taint of maya, and the Buddhist realization of shunyata. Philosophical conceptualization is incapable of describing this realization, for pure consciousness can only appear, even to the subtlest cognitive understanding, as nothing. For Samkhya, purusha is not nothing, but it is nothing that partakes of prakriti (which all awareness does).

Samkhya's unusual distinction between consciousness and what are ordinarily considered its functions and contents implies an operational view of purusha. Even though no properties can be predicated of purusha, the mind or intellect intuits the necessity of consciousness behind it, as it were. That is, the mind becomes aware that it is not itself pure consciousness. Since this awareness arises in individual minds, purusha is recognized by one or another egoity. Without being able to attribute qualities to purusha, it must therefore be treated philosophically as a plurality. Hence it is said that there are literally innumerable purushas, none of which have any distinguishing characteristics.

The Leibnizian law of the identity of indiscernibles cannot be applied to purusha, despite the philosophical temptation to do so, precisely because philosophy necessarily stops at the limit of prakriti. Purusha is outside space and time, and so is also beyond space-time identities. Since the minimum requirements of differentiation involve at least an indirect reference to either space or time, their negation in the concept of indiscernibility also involves such a reference, and cannot be applied to purusha. Even though Samkhya affirms a plurality of purushas, this stance is less the result of metaphysical certitude than of the limitations imposed by consistency of method. The plurality of purushas is the consequence of the limits of understanding.

Within the enormous and diverse history of Indian thought, the six darshanas viewed themselves and one another in two ways. Internally, each standpoint sought clarity, completeness and consistency without reference to other darshanas. Since, however, the darshanas were committed to the proposition that they were six separate and viable perspectives on the same reality, they readily drew upon one another's insights and terminology and forged mutually dependent relationships.

They were less concerned with declaring one another true or false than with understanding the value and limitations of each in respect to a complete realization of the ultimate and divine nature of things. Whilst some Western philosophers have pointed to the unprovable Indian presupposition that the heart of existence is divine, the darshanas reverse this standpoint by affirming that the core of reality is, almost definitionally, the only basis for thinking of the divine.

In other words, reality is the criterion of the divine, and no other standard can make philosophical sense of the sacred, much less give it a practical place in human psychology and ethics. In their later developments, the darshanas strengthened their internal conceptual structures and ethical architectonics by taking one another's positions as foils for self-clarification.

Earlier developments were absorbed into later understanding and exposition. Historically, Samkhya assimilated and redefined much of what had originally belonged to Nyaya and Vaishesika, and even Mimansa, only to find much of its terminology and psychology incorporated into Vedanta, the most trenchantly philosophical of the darshanas. At the same time, later Samkhya borrowed freely from Vedantin philosophical concepts to rethink its own philosophical difficulties.

Despite Samkhya's unique distinction between consciousness and awareness, which allowed it to preserve its fundamental dualism in the face of monistic arguments–and thereby avoid the metaphysical problems attending monistic views–it could not avoid one fundamental philosophical question: What is it to say that prakriti is dynamic because of the presence of purusha? To say that prakriti reflects the presence of purusha, or that purusha is reflected in prakriti, preserves a rigid distinction between the two, for neither an object reflected in a mirror nor the mirror is affected by the other. But Samkhya characterizes the ordinary human condition as one of suffering, which is the manifest expression of the condition of avidya, ignorance.

This condition arises because purusha falsely identifies with prakriti and its evolutes. Liberation, mukti, is the result of viveka, discrimination, which is the highest knowledge. Even though viveka might be equated with pure perception as the sakshi or Witness, the process of attaining it suggests either an intention on the part of purusha or a response on the part of prakriti, if not both. How then can purusha be said to have no relation, including no passive relation, to prakriti? Even Ishvarakrishna's enchanting metaphor of the dancer before the host of spectators does not answer the question, for there is a significant relationship between performer and audience.

Such questions are worthy of notice but are misplaced from the Samkhya standpoint. If philosophical understanding is inherently limited to the functions of the mind (which is an evolute of prakriti), it can encompass neither total awareness (purusha) nor the fact that both purusha and prakriti exist. This is the supreme and unanswerable mystery of Samkhya philosophy, the point at which Samkhya declares that questions must have an end. It is not, however, an unaskable or meaningless question. If its answer cannot be found in philosophy, that is because it is dissolved in mukti, freedom from ignorance, through perfect viveka, discrimination. In Samkhya as in Vedanta, philosophy ends where realization begins. Philosophy does not resolve the ultimate questions, even though it brings great clarity to cognition. Philosophy prepares, refines and orients the mind towards a significantly different activity, broadly called "meditation," the rigorous cultivation of clarity of discrimination and concentrated, pellucid insight.

The possibility of this is provided for by Samkhya metaphysics through its stress on the asymmetry between purusha and prakriti, despite their copresence. Prakriti depends on purusha, but purusha is independent of everything; purusha is pure consciousness, whilst prakriti is unself-conscious. Prakriti continues to evolve because individual selves in it do not realize that they are really purusha and, therefore, can separate themselves from prakriti, whilst there can never be complete annihilation of everything or of primordial matter.

Whereas Yoga accepted the postulates of Samkhya and also utilized its categories and classifications, all these being in accord with the experiences of developed yogins, there are significant divergences between Yoga and Samkhya. The oldest Yoga could have been agnostic in the sense implicit in the Rig Veda Hymn to Creation, but Patanjali's Yoga is distinctly theistic, diverging in this way from atheistic Samkhya. Whilst Samkhya is a speculative system, or at least a conceptual framework, Yoga is explicitly experiential and therefore linked to an established as well as evolving consensus among advanced yogins.

This is both illustrated and reinforced by the fact that whereas Samkhya maps out the inner world of disciplined ideation in terms of thirteen evolutes–buddhi, ahankara, manas and the ten indriyas–Patanjali's Yoga subsumes all these under chitta or consciousness, which is resilient, elastic and dynamic, including the known, the conceivable, the cosmic as well as the unknown. Whereas Samkhya is one of the most self-sufficient or closed systems, Yoga retains, as a term and in its philosophy, a conspicuously open texture which characterizes all Indian thought at its best. From the Vedic hymns to even contemporary discourse, it is always open-ended in reference to cosmic and human evolution, degrees of adeptship and levels of initiatory illumination.

It is ever seeing, reaching and aspiring, beyond the boundaries of the highest thought, volition and feeling; beyond worlds and rationalist systems and doctrinaire theologies; beyond the limits of inspired utterance as well as all languages and all possible modes of creative expression. Philosophy and mathematics, poetry and myth, idea and icon, are all invaluable aids to the image-making faculty, but they all must point beyond themselves, whilst they coalesce and collapse in the unfathomable depths of the Ineffable, before which the best minds and hearts must whisper neti neti, "not this, not that." There is only the Soundless Sound, the ceaseless AUM in Boundless Space and Eternal Duration.

Yoga

Almost nothing is known about the sage [Patanjali] who wrote the *Yoga Sutras*. The dating of his life has varied widely between the fourth century B.C.E. and the sixth century C.E., but the fourth century B.C.E. is the period noted for the appearance of aphoristic literature. Traditional Indian literature, especially the Padma Purana, includes brief references to Patanjali, indicating that he was born in Illavrita Varsha. Bharata Varsha is the ancient designation of Greater India as an integral part of Jambudvipa, the world as conceived in classical topography, but Illavrita Varsha is not one of its subdivisions.

It is an exalted realm inhabited by the gods and enlightened beings who have transcended even the rarefied celestial regions encompassed by the sevenfold Jambudvipa. Patanjali is said to be the son of Angira and Sati, to have married Lolupa, whom he discovered in the hollow of a tree on the northern slope of Mount Sumeru, and to have reduced the degenerate denizens of Bhotabhandra to ashes with fire from his mouth. Such legendary details conceal more than they reveal and suggest that Patanjali was a great Rishi who descended to earth in order to share the fruits of his wisdom with those who were ready to receive it.

Some commentators identify the author of the Yoga Sutras with the Patanjali who wrote the Mahabhashya or Great Commentary on Panini's famous treatise on Sanskrit grammar sometime between the third and first centuries B.C.E. Although several scholars have contended that internal evidence contradicts such an identification, others have not found this reasoning conclusive. King Bhoja, who wrote a well-known commentary in the tenth century, was inclined to ascribe both works to a single author, perhaps partly as a reaction to others who placed Patanjali several centuries C.E. owing to his alleged implicit criticisms of late Buddhist doctrines. A more venerable tradition, however, rejects this identification altogether and holds that the author of the Yoga Sutras lived long before the commentator on Panini. In this view, oblique references to Buddhist doctrines are actually allusions to modes of thought found in some Upanishads.

In addition to our lack of definite knowledge about Patanjali's life, confusion arises from contrasting appraisals of the Yoga Sutras itself. There is a strong consensus that the Yoga Sutras represents a masterly compendium of various Yoga practices which can be traced back through the Upanishads to the Vedas. Many forms of Yoga existed by the time this treatise was written, and Patanjali came at the end of a long and ancient line of yogins. In accord with the free-thinking tradition of shramanas, forest recluses and wandering mendicants, the ultimate vindication of the Yoga system is to be found in the lifelong experiences of its ardent votaries and exemplars. The Yoga Sutras constitutes a practitioner's manual, and has long been cherished as the pristine expression of Raja Yoga. The basic texts of Raja Yoga are Patanjali's Yoga Sutras, the Yogabhashya of Vyasa and the Tattvavaisharadi of Vachaspati Mishra. Hatha Yoga was formulated by Gorakshanatha, who lived around 1200 C.E. The main texts of this school are the

Goraksha Sutaka, the Nathayoga Pradipika of Yogindra of the fifteenth century, and the later Shivasamhita. Whereas Hatha Yoga stresses breath regulation and bodily discipline, Raja Yoga is essentially concerned with mind control, meditation and self-study.

The Yoga Sutras of Patanjali is universal in the manner of the Bhagavad Gita, including a diversity of standpoints whilst fusing Samkhya metaphysics with bhakti or self-surrender. There is room for differences of emphasis, but every diligent user of Patanjali's aphorisms is enabled to refine aspirations, clarify thoughts, strengthen efforts, and sharpen focus on essentials in spiritual self-discipline. Accommodating a variety of exercises–mind control, visualization, breath, posture, moral training–Patanjali brings together the best in differing approaches, providing an integrated discipline marked by moderation, flexibility and balance, as well as degrees of depth in meditative absorption.

The text eludes any simple classification within the vast resources of Indian sacred literature and a fortiori among the manifold scriptures of the world. Although it does not resist philosophical analysis in the way many mystical treatises do, it is primarily a practical aid to the quest for spiritual freedom, which transcends the concerns of theoretical clarification. Yet like any arcane science which necessarily pushes beyond the shifting boundaries of sensory experience, beyond conventional concepts of inductive reasoning and mundane reality, it reaffirms at every point its vital connection with the universal search for meaning and deliverance from bondage to shared illusions. It is a summons to systematic self-mastery which can aspire to the summits of gnosis.

The actual text as it has come down to the present may not be exactly what Patanjali penned. Perhaps he reformulated in terse aphoristic language crucial insights found in time-honoured but long-forgotten texts. Perhaps he borrowed terms and phrases from diverse schools of thought and training. References to breath control, pranayama, can be found in the oldest Upanishads, and the lineaments of systems of Yoga may be discerned in the Maitrayana, Shvetashvatara and Katha

Upanishads, and veiled instructions are given in the "Yoga" Upanishads–Yogatattva, Dhyanabindu, Hamsa, Amritanada, Shandilya, Varaha, Mandala Brahmana, Nadabindu and Yogakundali–though a leaning towards Samkhya metaphysics occurs only in the Maitrayana. The Mahabharata mentions the Samkhya and the Yoga as ancient systems of thought. Hiranyagarbha is traditionally regarded as the propounder of Yoga, just as Kapila is known as the original expounder of Samkhya. The Ahirbudhnya states that Hiranyagarbha disclosed the entire science of Yoga in two texts–the Nirodha Samhita and the Karma Samhita. The former treatise has been called the Yoganushasanam, and Patanjali also begins his work with the same term. He also stresses nirodha in the first section of his work.

In general, the affinities of the Yoga Sutras with the texts of Hiranyagarbha suggest that Patanjali was an adherent of the Hiranyagarbha school of Yoga, and yet his own manner of treatment of the subject is distinctive. His reliance upon the fundamental principles of Samkhya entitle him to be considered as also belonging to the Samkhya Yoga school. On the other hand, the significant variations of the later Samkhya of Ishvarakrishna from older traditions of proto-Sankhya point to the advantage of not subsuming the Yoga Sutras under broader systems. The author of Yuktidipika stresses that for Patanjali there are twelve capacities, unlike Ishvarakrishna's thirteen, that egoity is not a separate principle for Patanjali but is bound up with intellect and volition. Furthermore, Patanjali held that the subtle body is created anew with each embodiment and lasts only as long as a particular embodiment, and also that the capacities can only function from within. Altogether, Patanjali's work provides a unique synthesis of standpoints and is backed by the testimony of the accumulated wisdom derived from the experiences of many practitioners and earlier lineages of teachers.

Some scholars and commentators have speculated that Patanjali wrote only the first three padas of the Yoga Sutras, whilst the exceptionally short fourth pada was added later. Indeed, as early as the writings of King Bhoja, one verse in the fourth pada (IV. 16) was recognized as a line interpolated from Vyasa's seventh commentary in which he dissented from Vijnanavadin Buddhists. Other interpolations may have occurred even in the first three padas, such as III.22, which some classical commentators questioned. The fact that the third pada ends with the word iti ("thus," "so," usually indicating the end of a text), as it does at the end of the fourth pada, might suggest that the original contained only three books. However, the philosophical significance of the fourth pada is such that the coherence of the entire text need not be questioned on the basis of inconclusive speculations.

Al-Biruni translated into Arabic a book he called Kitab Patanjal (The Book of Patanjali), which he said was famous throughout India. Although his text has an aim similar to the Yoga Sutras and uses many of the same concepts, it is more theistic in its content and even has a slightly Sufi tone. It is not the text now known as the Yoga Sutras, but it may be a kind of paraphrase popular at the time, rather like the Dnyaneshwari, which stands both as an independent work and a helpful restatement of the Bhagavad Gita. The Kitab translated by al-Biruni illustrates the pervasive influence of Patanjali's work throughout the Indian subcontinent.

For the practical aspirant to inner tranquillity and spiritual realization, the recurring speculations of scholars and commentators, stimulated by the lack of exact historical information about the author and the text, are of secondary value. Whatever the precise details regarding the composition of the treatise as it has come down through the centuries, it is clearly an integrated whole, every verse of which is helpful not only for theoretical understanding but also for sustained practice.

The Yoga Sutras constitutes a complete text on meditation and is invaluable in that every sutra demands deep reflection and repeated application. Patanjali advocated less a doctrinaire method than a generous framework with which one can make experiments with truth, grow in comprehension and initiate progressive awakenings to the supernal reality of the Logos in the cosmos.

The word yoga is derived from the Sanskrit verbal root yuj, "to yoke" or "to join," related to the Latin jungere, "to join," "to unite." In its broadest usages it can mean addition in arithmetic; in astronomy it refers to the conjunction of stars and planets; in grammar it is the joining of letters and words. In Mimamsa philosophy it indicates the force of a sentence made up of united words, whilst in Nyaya logic it signifies the power of the parts taken together. In medicine it denotes the compounding of herbs and other substances.

In general, yoga and viyoga pertain to the processes of synthesis and analysis in both theoretical and applied sciences. Panini distinguishes between the root yuj in the sense of concentration (samadhi) and yujir in the sense of joining or connecting. Buddhists have used the term yoga to designate the withdrawal of the mind from all mental and sensory objects. Vaishesika philosophy means by yoga the concentrated attention to a single subject through mental abstraction from all contexts. Whereas the followers of Ramanuja use the term to depict the fervent aspiration to join one's ishtadeva or chosen deity, Vedanta chiefly uses the term to characterize the complete union of the human soul with the divine spirit, a connotation compatible with its use in Yoga philosophy. In addition, Patanjali uses the term yoga to refer to the deliberate cessation of all mental modifications.

Every method of self-mastery, the systematic removal of ignorance and the progressive realization of Truth, can be called yoga, but in its deepest sense it signifies the union of one's apparent and fugitive self with one's essential nature and true being, or the conscious union of the embodied self with the Supreme Spirit. The Maitrayana Upanishad states: "Carried along by the waves of the qualities darkened in his imagination, unstable, fickle, crippled, full of desires, vacillating, he enters into belief, believing I am he, this is mine, and he binds his self by his self as a bird with a net.

Therefore a man, being possessed of will, imagination and belief, is a slave, but he who is the opposite is free. For this reason let a man stand free from will, imagination and belief. This is the sign of liberty, this is the path that leads to brahman, this is the opening of the door, and through it he will go to the other shore of darkness."

Thus, yoga refers to the removal of bondage and the consequent attainment of true spiritual freedom. Whenever yoga goes beyond this and actually implies the fusion of an individual with his ideal, whether viewed as his real nature, his true self or the universal spirit, it is gnostic self-realization and universal self-consciousness, a self-sustaining state of serene enlightenment. Patanjali's metaphysical and epistemological debt to Samkhya is crucial to a proper comprehension of the Yoga Sutras, but his distinct stress on praxis rather than theoria shows a deep insight of his own into the phases and problems that are encountered by earnest practitioners of Yoga. His chief concern was to show how and by what means the spirit, trammelled in the world of matter, can withdraw completely from it and attain total emancipation by transforming matter into its original state and thus realize its own pristine nature. This applies at all levels of self-awakening, from the initial cessation of mental modifications, through degrees of meditative absorption, to the climactic experience of spiritual freedom.

Patanjali organized the Yoga Sutras into four padas or books which suggest his architectonic intent. Samadhi Pada, the first book, deals with concentration of mind (samadhi), without which no serious practice of Yoga is possible. Since samadhi is necessarily experiential, this pada explores the hindrances to and the practical steps needed to achieve alert quietude. Both restraint of the senses and of the discursive intellect are essential for samadhi. Having set forth what must be done to attain and maintain meditative absorption, the second book, Sadhana Pada, provides the method or means required to establish full concentration.

Any effort to subdue the tendency of the mind to become diffuse, fragmented or agitated demands a resolute, consistent and continuous practice of self-imposed, steadfast restraint, tapas, which cannot become stable without a commensurate disinterest in all phenomena. This relaxed disinterestedness, vairagya, has nothing to do with passive indifference, positive disgust, inert apathy or feeble-minded ennui as often experienced in the midst of desperation and tension in daily affairs. Those are really the self-protective responses of one who is captive to the pleasure-pain principle and is deeply vulnerable to the flux of events and the vicissitudes of fortune. Vairagya implies a conscious transcendence of the pleasure-pain principle through a radical reappraisal of expectations, memories and habits.

The pleasure-pain principle, dependent upon passivity, ignorance and servility for its operation, is replaced by a reality principle rooted in an active, noetic apprehension of psycho-spiritual causation. Only when this impersonal perspective is gained can the yogin safely begin to alter significantly his psycho-physical nature through breath control, pranayama, and other exercises.

The third book, Vibhuti Pada, considers complete meditative absorption, sanyama, its characteristics and consequences. Once calm, continuous attention is mastered, one can discover an even more transcendent mode of meditation which has no object of cognition whatsoever. Since levels of consciousness correspond to planes of being, to step behind the uttermost veil of consciousness is also to rise above all manifestations of matter. From that wholly transcendent standpoint beyond the ever-changing contrast between spirit and matter, one may choose any conceivable state of consciousness and, by implication, any possible material condition. Now the yogin becomes capable of tapping all the siddhis or theurgic powers. These prodigious mental and moral feats are indeed magical, although there is nothing miraculous or even supernatural about them.

They represent the refined capacities and exalted abilities of the perfected human being. Just as any person who has achieved proficiency in some specialized skill or knowledge should be careful to use it wisely and precisely, so too the yogin whose spiritual and mental powers may seem practically unlimited must not waste his energy or misuse his hard-won gifts. If he were to do so, he would risk getting entangled in worldly concerns in the myriad ways from which he had sought to free himself. Instead, the mind must be merged into the inmost spirit, the result of which is kaivalya, steadfast isolation or eventual emancipation from the bonds of illusion and the meretricious glamour of terrestrial existence.

In Kaivalya Pada, the fourth book which crowns the Yoga Sutras, Patanjali conveys the true nature of isolation or supreme spiritual freedom insofar as it is possible to do so in words. Since kaivalya is the term used for the sublime state of consciousness in which the enlightened soul has gone beyond the differentiating sense of "I am," it cannot be characterized in the conceptual languages that are dependent on the subject-object distinction. Isolation is not nothingness, nor is it a static condition. Patanjali throws light on this state of gnosis by providing a metaphysical and metapsychological explanation of cosmic and human intellection, the operation of karma and the deep-seated persistence of the tendency of self-limitation. By showing how the suppression of modifications of consciousness can enable it to realize its true nature as pure potential and master the lessons of manifested Nature, he intimates the immense potency of the highest meditations and the inscrutable purpose of cosmic selfhood.

The metapsychology of the Yoga Sutras bridges complex metaphysics and compelling ethics, creative transcendence and critical immanence, in an original, inspiring and penetrating style, whilst its aphoristic method leaves much unsaid, throwing aspirants back upon themselves with a powerful stimulus to self-testing and self-discovery. Despite his sophisticated use of Samkhya concepts and presuppositions, Patanjali's text has a universal appeal for all ardent aspirants to Raja Yoga. He conveys the vast spectrum of consciousness, diagnoses the common predicament of human bondage to mental ailments, and offers practical guidance on the arduous pathway of lifelong contemplation that could lead to the summit of self-mastery and spiritual freedom.

Introduction

Here we have a lecture series dealing with the systems of Indian Philosophy and delivered by V. R. Gandhi in 1894 at

Chicago. These lectures are important as much because they deal with the systems of Indian Philosophy as because V. R. Gandhi delivered them. For V. R. Gandhi (who was born in 1864 and died young in 1901) was one of the extraordinary Indians of his time. He was a born Jaina and (what is more noteworthy) a convinced Jaina, and it was as representative of the Jaina sect that he took part in the Parliament of Religions held at Chicago in 1893 (better known to most of us on account of Swami Vivekananda's participation in it).

But few Jainas before and after him would equal him in their capacity to make the Jaina positions comprehensible to a non-Jaina audience and in their capacity to adopt a most non-sectarian approach while dealing with a problem. Gandhi's many lectures meant to undertake an exposition of the various aspects of Jainism (and his article "Philosophy and Psychology of the Jains" published in Mind Vol. I, No. 4)-most of them available to us in the collection published under the title "The Jaina Philosophy"-can well form for those who know English a best introduction to this branch of studies in Indian culture. Particularly noteworthy in this connection are the lectures (delivered in England) dealing with the Jaina doctrine of *Karma*.

The verbatim notes of these lectures-which were in possession of H. Warren and were probably taken down by himself-were later on published under the title "The *Karma* Philosophy". V. Glasenapp, the recognized Western authority on Jainism in general and the Jaina doctrine of *Karma* in particular, duly acknowledges his indebtedness to these lectures of Gandhi which even today remain an independent source of enlightenment on the subject in spite of the Gedrman scholar's doctoral dissertation devoted to the same.

The "doctrine of *Karma*", subscribed to by the Vedicists, *Buddhi*sts, Jainas and numerous other religious sects of India, holds a crucial importance in the development of the characteristic ethical notions of the ancient Indians, and the Jaina version of it is illuminating in more ways than one. It is really a pity that even so lucid an exposition of the Jaina doctrine of *Karma* as was undertaken by Gandhi remains unread even by those who otherwise evince sincere and serious interest in the problems of Indian ethics. Of course, in order to derive best advantage out of Gandhi's writings things will have to be looked from Gandhi's standpoint. There are times when Gandhi speaks as a Jaina, times when he speaks as a Hindu, times when he speaks as an Indian, and times when he speaks as a plain man.

While speaking as a Jaina, a Hindu, or an Indian, Gandhi is in most cases positive in his assertions, that is, he mostly brings to the fore the merits of the case he is advocating; but occasionally he is forced to come out sometimes sharply enough against what he considers to be a gross misunderstanding of his case on somebody's part. He is bitterest in his condemnation of the Christian missionaries, come to India from abroad to propagate their cult. But his motives in doing so are extremely mixed. Gandhi is against the Christian missionaries because the latter consider the Hindu to be ethically degraded. Now Gandhi would not answer this slander by talking ill of Christians en masse, not only because he had nothing, but praise for what he considered to be Christ's true teaching, but also because he had come to cultivate warm friendship with a vast number of noble-minded Christians both in England and in America. Gandhi therefore took care to distinguish between the ordinary Christian residing in England or America and the Christian missionaries who come to India from abroad; in his lectures like "India's Message to America" and "Impressions of America" he paid handsome tributes to the former, in those like "Have Christian Missions to India been Successful?" he cursed the latter.

As an Indian Gandhi was painfully conscious of his country's dependent status as also of the economic exploitation this country was subjected to, but his observations on these matters are mostly in the form of obiter dicta. For example, in the course of his "India's Message to America" he makes bold to say: "You know, my brothers and sisters, that we are not an independent nation; we are subjects of Her Gracious Majesty Queen Victoria, the 'defender of the faith', but if we were a nation in all that that name implies, with our own government and our own rulers, with our laws and institutions controlled by us free and independent, I affirm that we should seek to establish and for ever maintain peaceful relations with all the nations of the world" (The Jaina Philosophy, p. 264). A still more revealing passage-occurring in "Have Christian Missions to India been Successful?"-runs as follows: "Ladies and gentlemen, you have heard all yours lives from your missionaries who claim to be the messengers of God how ugly, wretched, immoral, and vile the heathen Indians are;... but did you ever hear from these missionaries-the messengers of love to all mankind-of the tyrannies that are perpetrated over the Hindus in India? Government has abolished duties on fine dry goods from Liverpool and Manchester for the purpose of finding a good market in India and has levied a 200 per cent tax on the manufacture of salt in India to maintain a costly government. Did they ever tell you about all such things?

If they have not, whose messengers you will call these people, who always side with tyranny, who throw their cloak of hypocritical religion over murderers and all sorts of criminals who happen to belong to their religion or to their country?" (The Jaina Philosophy, pp. 85-86). Thus Gandhi dreamt of an India politically and economically independent but he was intelligent enough to see that there was no immediate prospect of his dream coming true. On the other hand, what might be called India's "religious independence" was a glowing reality before Gandhi's eyes and he was extremely anxious lest this too should gradually become extinct. Hence his tirade against the Christian missionaries. Let us however not forget that Gandhi's chief weapon in the struggle for what was in his eyes his country's "religious survival" was positive rather than negative. That is to say, Gandhi was interested not so much in saying things against the Christian missionaries as in saying things in favor of India's cultural heritage, a heritage to which his own Jaina community had made no mean contribution.

This background to Gandhi's activities explains, why he always spoke with the zeal of a missionary. But significantly enough, in Gandhi's mental make-up there was also a scholarly side and the best literary specimens, where he comes out as a beautiful blend of the missionary, and the scholar are his lectures pertaining to Jainism-particularly those related to the Jaina doctrine of *Karma*. A specimen belonging to the same group is his present lecture-series dealing with the systems of Indian philosophy. However, this series has certain specific features of its own, and it is to these that we turn our attention next.

The task of interpreting the systems of Indian Philosophy is beset with two sets of problems, one having to do with the nature of the subject-matter in question and the other with what happens to be the general standpoint of the interpreter concerned. To take the two sets one by one. The major part of India's philosophical literature is in Sanskrit, some in Prakrit and some in Pali; and almost no texts that claim attention in this connection are a modern composition. Thus a student of Indian philosophy has not only to master a language like Sanskrit (preferably, Prakrit and Pali as well) but he has also to learn the art of placing himself in the position of an ancient or a medieval Indian.

It is only after fulfilling these two rather irksome requirements that one would find it possible to rightly understand what a particular system of Indian philosophy says on this or that problem it has cared to investigate. And then comes the question of offering interpretations to what has been taught by a system of Indian Philosophy, interpretations that are bound to differ in case they happen to be offered by students whose own ideological affiliations are mutually different. Of course, the ideological affiliation of an interpreter of Indian Philosophy (for that matter, of any philosophy whatsoever) need not bear a recognized 'label' but it should be something precisely definable nevertheless. For example, the general standpoint of Radhakrishnan (and of those numerous prominent Indian authors who have followed his lead) can rightly be called *Advaita Vedantic*, but it will be somewhat difficult to give a name to the general standpoint of a Max Muller or a Deussen.

But both Max Muller and Deussen were good Christians deeply in sympathy with Kant, and the fact is largely responsible for the way they have handled the problems of Indian

Philosophy. Certainly, a Western movement for the study of Indian Philosophy headed by persons like Max Muller and Deussen, could not but present the Advaita Vedanta of Sankara in the most favorable light, and judge each and every other systems of Indian Philosophy on the basis of the distance that separates it from this Advaita Vedanta, a procedure essentially the same as was subsequently followed by Radhakrishnan and others in India.

This circumstance is a good deal responsible for the somewhat lop-sided development of the studies related to Indian Philosophy that have been conducted in the West and in India in the course of past hundred years or so. Gandhi's keen eyes could see the danger inherent in the situation, as should be evident from the following comment he made (in his article published in Mind) by way of taking mild exception to a statement occurring in the Prospectus of the newly founded journal that was to acquire a big name afterwards: "This statement seems to whisper in my ears that Hindu metaphysics has not been able to offer the right solution of the various intricate problems of life that are staring in the face of the Western thinkers. By "Hindu" is meant, of course, the special phase of Vedanta philosophy that has been presented to the people of West during the last four years. I am glad that the truth in Vedanta has come to the shores of this country.

It would have been much better if the whole truth lying back of the different sectarian systems of India had been presented, so that a complete instead of a partial view of India's wisdom might have satisfied the craving of deep students." (The Jaina Philosophy, p. 14). Be that as it may, the systems of Indian Philosophy can be fruitfully studied also from a Western standpoint different from that of Kant and from an Indian standpoint different form that of Sankara. Nay, it is doubtless desirable that these systems be studied from the various standpoints that dominate the Western philosophical scene as also from those that dominate the Indian philosophical scene. Gandhi's present lectures on the systems of Indian Philosophy are important inasmuch as they give us an idea of how a liberal Jaina looks at-and places before an American audience-the philosophical heritage of his motherland. Gandhi well realized that grounding in Sanskrit is indispensable for one seeking to know something of India's past glory. That is why he once argues: "The many learned missionary gentlemen who have written or who have exhausted their oratory power in denouncing India, can only prove their claim to be an authority when they show their knowledge of the Hindu religion, and this can only be proven by their knowledge of Sanskrit. When they can converse with me in this language I Shal consider their words worthy of consideration and not before". ("Have Christian Missions, etc.", The Jaina Philosophy, p. 86)

Of course, Gandhi was not only not blind to the existence of Western Sanskritists but was himself a personal friend of good many of them; (what he was there criticizing was the ignorant debunking of things Indian on the part of the Christian missionaries come from abroad). Not only that, he actually made best use of the English translations done by Western scholars of the Sanskrit, Prakrit and Pali texts, though when need arose, he would prepare his own English version of an Indian text passage that was in no way inferior to that of the best translators of those days.

As a matter of fact, Gandhi's general mastery over English language was strikingly perfect. However, a thorough grounding in Sanskrit and a good command over English would not have sufficed for Gandhi's need; what he above all required was a capacity to grasp the spirit of the teaching imparted by an ancient Indian text, he took up for study. And with this capacity too Gandhi was endowed in good measure. A ringing confirmation of this comes from his present lectures on the systems of Indian Philosophy, where we find him taking great pains to tell us just, what a Samkhya Philosopher, a Yoga Philosopher, a *Naya-Vaisesika* Philosopher, a Vedanta Philosopher or a *Buddhi*st Philosopher has to say on this or that question.

Of course, the very fact that Gandhi chooses to discuss certain topics and not others in the course of his treatment of a particular system of Indian Philosophy betrays his own likes and dislikes; the more so is the case with the critical remarks he now and then passes against a non-Jaina system. But that has to be the feature of all principled exposition of the tenets of Indian Philosophy (for that matter, of any philosophy whatsoever); and Gandhi was certainly a man of principles. What we are emphasizing is that Gandhi's own ideological affiliation did no prevent him from making maximum effort to get at the heart of the various positions developed by the various non Jaina systems of Indian Philosophy. In his lecture on Jainism-which is the last lecture in the present series-Gandhi enumerates what he considers to be the four questions basic to all philosophical investigation; they are:

(i) What is the nature of the universe?

(ii) What is the nature of God?

(iii) What is the nature and what the destiny of soul?

(iv) What are the laws of the soul's life?

[the questions (iii) and (iv) are closely related, the former inquiring about the general nature of a soul, its bondage and its liberation, the later *inquiring* about the functioning of the "law of *Karma*"].

And his exposition of Jainism is in the form of a discussion of the Jaina answer to these four questions. In the case of the rest of the systems there is no ordered treatment of these questions, but there too Gandhi is always taking up one or another from among these very questions (which is but to be expected in view of Gandhi's understanding of what constitutes a philosophical investigation being what it is). And it should not be difficult for an intelligent reader to make out for himself how this or that system differs from Jainism on this or that question. But Gandhi, almost totally unmindful of this difference, continues his painstaking works of exposition. As for the points of criticism occasionally raised against a non-Jaina system they seem to have been balanced by an occasionally showered praise. In any case, Gandhi is not obsessed by the fact that each of the non-Jaina systems considered by him differs from Jainism more or less sharply on some questions or others.

Let us now take critical note of the facts about Indian Philosophy that Gandhi thought fit to convey to his American audience and of his manner of doing so Gandhi has taken up for consideration the following systems: *Samkhya*, Yoga, *Naya* (and *Vaisesika*). *Mimamsa, Vedanta*, *Buddhism* and *Jainism*. And it will be convenient and useful for us to discuss his treatment of these systems one by one.

1. Samkhya

Gandhi bases his account of the Samkhya system on the version of it that we find in the Samkhya *Sutras* (a version not essentially different from that found in the Samkhya Karika and one entitled to be treated as 'Classical Samkhya'). Students of Indian Philosophy attach importance to the Samkhya system for diverse-nay, mutually opposite-reasons. Those inclined to favour idealism (if the Advaita Vedanta type, say) emphasize the fact that according to Samkhya the world of day-to-day experience (in it capacity as an evaluate of *Prakrit*) is real to a soul-in-bondage (*i.e.* a soul-under-ignorance) but unreal to an emancipated (*i.e.* an enlightened) soul; those inclined to favour realism emphasize the fact that according to Samkhya *prakriti*, the root-cause of the world of day-to-day experience, is a reality co-eternal with the multiplicity of souls.

As a matter of fact, the Samkhya philosopher's position on the question is considerably obscure, it being really difficult to make out as to what he precisely means by his thesis that *prakriti* evolves itself in the form of the world of day-to-day experience for a soul that is in bondage while it ceases to do so for a soul that is emancipated. With this obscurity in the background we can easily follow Gandhi's account of the Samkhya system. Gandhi gives prominence to the Samkhya philosopher's contention that the world of day-to-day experience evolved out of *prakriti* is not an illusory appearance and that the souls are many in number, a contention directed against two fundamental theses of Advaita Vedanta. But he raises pointed objection against the Samkhya position that *Buddhi* ('intellect' in Gandhi's translation) is a product of *prakriti* (which in turn is a physical entity) while *ahankara* ('self-consciousness' in Gandhi ji's translation) is a product of *Buddhi*. The functions that the Samkhya philosopher assigns to *Buddhi* and *ahankara* will be assigned to soul by Gandhi (rather by the Jaina philosopher) and the latter must have noted that the former's way of speaking paves the way for the Advaita Vedantist's dismissal of a soul's individuality as an illusory appearance. For *Buddhi* and *ahankara* represent the essence of an individual's individuality, and if they have nothing to do with soul the conclusion certainly follows that soul has nothing to do with an individual's individuality; and this conclusion couple with the thesis that all physical phenomena whatsoever are illusory naturally leads to the Advaita Vedanta position that the sole existing reality is one soul.

Of course, Gandhi must have also realized that the functions attributed by the *Samkhya* philosopher to *Buddhi* and *ahankara* cannot be the functions of a physical entity (as *Buddhi* and *ahankara* allegedly are), for to concede that possibility will mean embracing materialism. Be that as it may, Gandhi made an honest attempt to place before his audience the picture of an Indian system of philosophy that is partly idealist, partly realist, partly materialist. And if it is the realistic aspects of the Samkhya teaching that chiefly received Gandhi's attention it is not because Gandhi was himself a realist but because the 'classical Samkhya' is actually a realistic system of philosophy on the whole. One more point. Gandhi well observed that in an Indian system of philosophy the metaphysical and ethic-religious matters invariably go hand in hand, but he also knew that the importance attached to these two in different systems is differently proportioned.

And consequently in his exposition of a system of Indian Philosophy Gandhi would endeavour to remain loyal to the spirit of the original in this respect. Thus he treated Samkhya as a philosophical system chiefly devoted to theoretical problems while touching upon the problems of practice as well; (on the contrary, he treated the *Yoga* of *Patanjali* as a philosophical system chiefly devoted to practical problems while touching upon the problems of theory as well). That is why Gandhi begins his lecture on Samkhya by telling us that the Samkhya philosopher aims at a cessation of the threefold miseries while in the course of his exposition he incidentally tells us as to what according to the Samkhya philosopher is the nature of *moksa* and what the means of attaining it, for the rest his concern is with the metaphysical tenets of the Samkhya system.

2. Yoga

Gandhi rightly noted that the *Yoga* system of philosophy-more properly, the system of philosophy propounded by *Pantanjali* in his *Yoga Sutras*-differs but little from Samkhya so far as theoretical questions are concerned; what distinguishes Yoga is its over-all preoccupation with practical matters. Hence we find Gandhi too almost exclusively discussing practical matters throughout his lecture on Yoga. But the practical matters taken into consideration by the Yoga system are of a somewhat peculiar nature.

The Yoga philosopher (rather the Yoga adept) aims at developing the capacity to concentrate his mind on one subject of the exclusion of everything else-and ultimately to concentrate it on 'nothing'. A rough equivalent for 'concentration of mind' is 'cessation of mental modifications (*Skt. Citta vrtti-nirodha*)' and whatever theoretical problems interest a Yoga philosopher mostly arise in the course of his inquiry into the precise nature of *citta, citta-vrtti* and *citta-vrtti-nirodha*. For the rest he is busy discussing the practical measures to be devised in order to develop the capacity for 'concentration of mind' (or discussing the miraculous capacities that a practicing *yogi* allegedly comes to acquire). Gandhi's exposition of *Yoga* therefore begins with a brief account of *citta, citta-vrtti* and *citta-vrtti-nirodha*; then is considers the nature of the eight *yogangas* (or 'means of yoga'-*i.e.*, means for developing the capacity for concentration of mind), and lastly the miraculous capacities that one allegedly comes to acquire as a result of concentrating one's mind on this object or that.

Now the first two *yogangas* happen to be *yam* and *niYams* (in Gandhi's translation 'forbearances' and 'observances') and the various sub-species of them happen to be various virtues of character. Thus the five *yams* are 'abstaining from killing (*ahimsa*)', 'abstaining from falsehood (*satya*)', 'abstaining from theft (*asteya*)', 'austerity (tapas)', 'study (*svadhyaya*)' and 'resignation to God (*Isvarapranidhana*)'. Hence the consideration of these two *yogangas* provided Gandhi a good opportunity to express his views on a number of ethical questions. Of course, in his exposition Gandhi did not want to deviate from what was actually said or implied in the *Yoga* writings; but when he found that a particular position adopted by the *Yoga* philosopher was not worth dilating upon he simply mentioned it and passed on.

This attitude becomes particularly striking in the later parts of his exposition-that is, in the course of his exposition of the remaining six *yogangas* and of the miraculous capacities allegedly acquired by a practicing *yogin*. In these parts we are able to know a good deal as to what the Yoga philosopher has to say on the questions under consideration but pretty little as to what Gandhi himself feels about the matter. But one thing is certain. In his own way Gandhi was thoroughly convinced that as a result of controlled 'concentration of mind' (and the allied yoga exercises) one can come to acquire supra-normal capacities of body and mind; this becomes clear not only from the occasional comments made by him in the course of his present lecture on Yoga philosophy but also from his numerous other lectures on the subject of yoga which were later on published under the title 'The Yoga Philosophy'. Perhaps, Gandhi would not therefore endorse the following stricture passed by Max Muller against that part of the Yoga *Sutras* where the miraculous powers allegedly acquired by a practicing *yogi* are enumerated: "... we get more and more into superstitions, by no means without parallels in other countries, but for all that, superstitions which have little claim on the attention of the philosopher, however interesting they may appear to pathologist", (The six systems of Indian philosophy, p. 351).

But then Max Muller had himself gone on to add; "These matters, though trivial, could not be passed over, whether we accept them as hallucinations to which, as we know, our thinking organ (organs?) are liable, or whether we try essential part on *yoga* philosophy and it is certainly noteworthy even from a philosophical point of view, that we find such vague and incredible statements side by side with the specimens of the most exact reasoning and careful observation" (Ibid., p. 352) Moreover, the acquisition of miraculous capacities was not considered even by Gandhi to be the true aim of yoga practice; for in his eyes this aim was 'self-culture' as he understood it.

3. Naya (and Vaisesika)

For reasons partly technical and partly ideological the *Naya-Vaisesika* system yet remains 'under-studied' by the students of Indian Philosophy-Indian as well as Western. On account of their logical rigor-as also on account of their highly evolved technical terminology-even the elementary *Naya-Vaisesika* texts are tough enough to scare the novice. Another reason for the comparative neglect of the system lies in the content of its teaching.

The *Naya-Vaisesika* philosophy is a type of empirical realism and as such it is opposed to the transcendental idealism of Advaita Vedanta-the system patronized by a majority of scholars working the field of Indian philosophy. Max Muller's attitude was typical. "While in the systems hitherto examined," he says, "particularly in the *Vedanta*, *Samkhya* and *Yoga*, there runs a strong religious and even poetical vein, we now come to two systems, *Naya* and *Vaisesika*, which are very dry and unimaginative,... businesslike exposition of what can be known, either of the world which surrounds us or of the world within..." (The Six Systems, p. 362). Gandhi, who was himself a man of deeply religious temperament, and who must have been alive to the fact that the *Naya-Vaisesika* system pays scant heed to the problems of ethics and religion, could not ditto Max Muller's sweeping condemnation of the system, not only because the condemnation was so sweeping but also because Gandhi's own general philosophical standpoint was realistic rather than idealistic.

But as things stood, Gandhi did not think it worthwhile to say much (maybe he had not think it worthwhile to say much (maybe he had not much to say) about the philosophical teachings of the *Naya-Vaisesika* system, and what we have from his pen is a barest outline of the sixteen topics (technically called *Padarthas*) whose consideration exhausts what may be called the *Naya* philosophy and of the seven categories (again, technically called *Padarthas*) whose consideration exhausts what may be called a *Vaisesika* philosophy.

4. Mimamsa

Gandhi did not consider *Mimamsa* to be a system of philosophy but a system of ritualism, and that is why he just takes note of it and then passes on to the system to be taken up next. As a matter of fact, *Mimamsa* is both a system of philosophy and a system of ritualism. But the philosophical literature emanating from the *Mimamsa* school belongs to the same broad category (and broadly presents the same type of difficulties before a student) as does that emanating from the *Naya-Vaisesika* school. Nay, a serious study of the *Naya-Vaisesika* philosophy is impossible without a serious study of the *Mimamsa* philosophy (just as it is impossible without a serious study of the *Buddhi*st philosophy as expounded by the school of Dinnaga and Dharmakirti). Be that as it may, we too take leave of *Mimamsa* and proceed on the Vedanta.

5. Vedanta

Gandhi's account of the Vedanta philosophy is most illuminating and for various reasons. Neither in the case of *Samkhya-Yoga*, nor in that of Naya-*Vaisesika* (nor in that of *Mimamsa*) did Gandhi encounter strong contemporary champions, but a good part of India's *Hindu* populace happens to be the adherent of one *Vedanta* sect or another (and a majority of scholars working in the field of Indian philosophy happen to be the sympathizers of *Advaita Vedanta*). Gandhi therefore thought it necessary to carefully analyze the respective philosophical standpoints of Sankar-the chief advocate of *Advaita Vedanta*-and Ramanuja-the chief advocate of *Visistadvaita Vedanta*-, devoting relatively much greater attention to the former. And by way of introducing his subject he quoted long passages from the famous *Chandogya Upanisad* dialogue between Uddalaka Aruni and his son Svetaketu. We are thus enabled to work out for ourselves of comparative estimate of the old-*Upanisadic* teaching, Sankara's teaching and Ramanuja's teaching on the fundamental questions of philosophy. In the course of his exposition of Sankara's philosophy Gandhi explicitly touches upon the problem of the relation in which this philosophy stands to the teaching contained in the old *Upanisads*. He rightly points out that Sankara's followers with their distinction between 'lower' and 'higher' truths find no difficulty in both accepting and repudiating the teaching of old *Upanishads* which seldom lend clear support to the idealist-illusionist philosophy of Sankara. As a matter of fact, in Gandhi's present lecture-series most of such remarks as can be construed as critical-remarks that are certainly few and far between-are concentrated in the part concerned with the exposition of Sankara's philosophy.

6. Buddhism

The last non-Jaina system of philosophy considered by Gandhi is *Buddhi*sm. But here the exposition of the *Buddhi*st philosophy is preceded by a summary narration of Budda's life-story. The decision of include the biographical portion seems to have been a result of second thoughts but it has been well executed; for we are thereby assisted in forming a graphic idea of what it was in Buddha's life-activities that Gandhi admired most. In his exposition of the *Buddhi*st philosophy Gandhi confines himself to Southern *Buddhi*sm (*i.e.*, the *Theravada* branch of *Hinayana Buddhi*sm). Now in the philosophical literature of Southern *Buddhi*sm much attention has been devoted to the ethico-religious problems and comparatively little to the metaphysical ones.

The same is the case with Gandhi's account of the *Buddhi*st philosophy. For we are here given an account of the fourfold 'noble truths', fthe seven 'jewels' of the *Buddhi*st law, the *Buddhi*st notion of nirvana, the *Buddhi*st understanding of the 'law of *Karma*', and such other ethico-religious topics, but the doctrine of 'five *skandhas* (along with its corollary, the doctrine of 'no soul')-the only metaphysical doctrine considered-is introduced as a sort of side-issue while dealing with the first 'noble truth'. The only place where Gandhi pointedly raises objection against a *Buddhi*st position is revealing. For he feels that Buddha's acceptance of the 'law of *Karma*' is incompatible with the latter's denial of 'soul'. Now irrespective of whether this objection of Gandhi is valid or not it is definitely indicative of his repeatedly asserted conviction that an ethics in order to be sound must be based on a sound metaphysics.

7. Jainism

Last of all Gandhi takes up the Jains system of philosophy, a system he himself espouses. As noted earlier, it is in this connection that Gandhi enumerates the four questions regarded by him as basic to all philosophical investigation. The questions are:

(1) What is the nature of the universe?

(2) What is the nature of God?

(3) What is the nature and what the destiny of soul?

(4) What are the laws of the soul's life?

Gandhi's account of the Jaina answer to these four questions is worthy of most serious consideration. For here we have a fine illustration of Gandhi's inexhaustible capacity to make the Jaina positions comprehensible to a non-Jaina audience-and a non-Jaina Western audience at that). Gandhi's 'four questions' clearly prove that his understanding of what constitutes a philosophical investigation was truly all-comprehensive. Thus he would expect a philosophical system to touch upon the problems of metaphysics, psychology, ethics, as well as religion. Of course, Gandhi knew (and the present lecture-series is an evidence thereof) that not all-philosophical systems are equally interested in discussing these various generic types of problems, but he was convinced-perhaps, rightly that neglect of any of these types of problems on the part of a philosophical findings.

It is hoped that this preliminary introduction to Gandhi's lecture-series on the systems of Indian Philosophy will help the reader in viewing it in a proper perspective.

The present edition of Gandhi's lecture-series is prepared on the basis of his own manuscript of it that is in the possession of Shri Mahavir Jain *Vidyalaya*, Bombay. However, this manuscript does not contain anything on Jainism. But the lecture (with the title 'Jainism') published on pp. 41-60 of The Jaina Philosophy begins by mentioning that it is the last lecture of some lecture series; from this we have surmised that here is the lecture on Jainism that belongs to our lecture-series (which too need in the form of its last member a lecture on Jainism). Maybe our surmise is wrong but most probably it is not. Again, we learn from The Universalist Messenger, Chicago, February 10, 1984 (quoted at the end of the 'Selected Speeches of Shri Virchand Raghavji Gandhi' published in May 1964 in the form of 'Shri Vallabhsuri Jaina Literature Series, No. 10') : "The series of lectures on Oriental philosophy given by Mr. Virchand R. Gandhi every Monday evening at the residence of Mr. Chas. Howard, 6558 Stewart Boulevard, are growing more and more interesting. The subject philosophy." This (along with the fact that the first blank page of our manuscript carries the address '6558, Stewart Avenue, Englewood III)' is the basis of our surmising that our lecture-series was delivered at Chicago in 1984. Here again our surmise might possibly be wrong but most probably it is not.

Mistakes occurring in the manuscript that are obviously the slips of pen have been corrected by us without making mention of the fact, but the places where a mistake is just suspected or where the manuscript is not legible have been duly noted. The division of a lecture into sections and of a section into paragraphs (as also the titling of sections) has been undertaken by as with view to facilitating the reader's comprehension and Yoga Gandhi closely follows certain texts of the systems; hence at appropriate places a precise reference to the relevant passages from these texts has been made by us in the form of footnotes.

In the case of *Buddhi*sm, similar reference has been made to a few passages from the *Abhidhammathasangaho*-a standard philosophical manual of *Theravada Buddhism*-; but this does not amount to claiming that it is this text that has been used by Gandhi. (The lectures on *Naya* and Vedanta are a few independent footnotes of our which seek either to elucidate of to complete or to criticize a remark made by Gandhi; (These are not footnotes given by Gandhi himself).

Following Gandhi's practice, no diacritical marks have been used in the Roman version of Indian proper names.

However, since the technical terms of Indian philosophy, when written in Roman without diacritical marks, are likely to be misunderstood they have been given in *Devanagari*; (this too is in most cases a practice also of Gandhi-who however uses for the purpose the Gujarati script rather than Devanagri).

2

The Samkhya Philosophy

1. We begin this evening with the *Samkhya* philosophy. Kapila, the reputed author of this philosophy was probably a Brahmin, Though nothing is known about him. He is the supposed author of two works-the original *Samkhya Sutras* called (*Samkhya Pravachan*) and a shorter work called (*Tatvsmas*). The *Samkhya* philosophy together *with Yoga, Naya, Vaisheshika, Mimamsa and Vedanta* nominally accepts *Veda* as its guide. It is the Philosophy of (*Samkhya*), *i.e.* enumeration or analysis of the Universe. Sir Monier Williams calls it by the name of synthetic enumeration. Sir William Jones calls it the Numeral Philosophy. It has been partly compared with the metaphysics of Pythagoras, partly in its *Yoga* with the system of Zeno. Others compare it with that of Berkeley.
2. It starts with the proposition that the world is full of miseries of three kinds-the three kinds of miseries:
   (1) (*Adhyatmic*) due to one's self,
   (2) (*Adhibhotic*) due to the products of elements and
   (3) (*Adhidaevic*) due to supernatural causes-and that the complete cessation of pain of theses three kinds is the complete end and object of man. (*Trividhasya adhyatmic Adhibhotic, Adhidaevic, roopsay, dukhsay, atyantnivriti atyantPurushrth.*)

   This doctrine of *Samkhya* is similar to the tenets held by the Buddhists whose main doctrine is that the world is full of miseries. This is also the starting point of

Spinoza. In his work 'The Improvement of the Understanding' he says: "After experience had taught me that all the usual surroundings of social life are vain and facile seeing that none of the objects of my fears contain in themselves anything either good or bad, except in so far as the mind is affected by them, I finally resolved to inquire whether there might be some real good which the discovery and attainment would enable me to enjoy continuous, supreme and unending happiness." That is his starting point, just the starting point where the *Samkhya* starts. He goes on to say: "I thus perceived that I was in a state of peril and I compelled myself to seek with all my strength for a remedy, however uncertain it might be, as a sick man struggling with a deadly disease when he sees that death will surely be upon him unless a remedy be found, is compelled to seek such a remedy with all his strength, in as much as his whole hope lies therein. All the objects pursued by the multitude not only bring no remedy that tends to preserve our being, but even act as hindrance, causing the death not seldom of those who are possessed by them." He continues: "All these evils seem to have arisen from the fact that our happiness or unhappiness has been made the mere creature of the thing that we happen to be loving. When a thing is not loved, no envy if another bears it away, no fear, no hate; yes, in a world no tumult of soul. These things all come from loving that which perishes, such as the objects of which I have spoken. But love towards a thing eternal feasts the mind with joy alone, nor hath sadness any part therein. Hence this is to be prized above all and to be sought for with all our might."

3. How was such a theory invented? In the West it has always been the case that the peculiar circumstances of the philosopher's life lead him into a peculiar belief, in the East the calm and quiet scenery and bountiful nature lead him to patiently inquire into the mysteries of the universe. Their contemporaries judge them from

a false vantage ground. Spinoza in his owns age was denounced as a atheist, profane person, monster. Long afterwards however his works were re-discovered, greedily read, and admired by great poets like Goethe and by ardent and even romantic philosophers like Schelling. The *Samkhya* system too was considered by its commentators atheistic. But the present generation looks charitably upon it and tries to see some if not all-eternal truths in it.

4. I told you in the beginning that the *Samkhya* starts with the proposition that the world is full of miseries of three kinds. These are the results of the properties of matter (*Prakriti*) and not of its correlate intelligence of consciousness (*Purush*). Matter is eternal and co-existent with spirit. It was never in a state of non-being but always in a state of constant change, it is subtle and insentient. According to this view, *Prakriti* existed before the evolution of the universe and will continue so to exist for ever, but with time it has so much been changed that the unemancipated (*Atma*) (soul) is but ill able to comprehend its nature. It has lost its original state and has become earthy. In other words, *Prakriti* has assumed diverse shapes both gross and subtle.
5. Kapila's theory is strictly a theory of evolution. He says: (*Navstuno vstusidhi*) — A thing is not made out of nothings. *Avastunobhavat vastusidhirbhavotpatirnav sambhavti* It is not possible that out of nothing, *i.e.* an entity should arise. (*Yadyabhavat bhavotpatistarhi karan rupan karyai drishyat iti jagtopyavastusvanlllll*) If an entity were to arise out of a non-entity, then since the character of a cause is visible in its product the world also will be unreal. When the *Vedantist*-the monist or the idealist-tell Kapila, 'Let the world too be unreal, what harm is that to us?', he replies : *Abadhat adushtkaran janyatvach navstuutvama*-The world is not unreal because these is no fact contradictory to its reality and because it is not the false result of depraved causes (leading to a belief in what ought not to be

believed). (*Ahuktao rajatmiti gyanai naidan rajatmiti gyanat naidan rajatbadh na chatr naidan bbhavroopan jagditi ksyapi gyanan yain bhavroopbadh syat)* When there is the notion in regard to a shell of a pearl-oyster (which sometimes glitters like silver) that it is silver, its being silver is contradicted by the subsequent and more correct cognition that this is not silver. But in the case in question-that of the world regarded as a reality, no one ever has the cognition "this world is not in the shape of an entity", by which cognition if any one ever really had such its being an entity might be opposed. (*Dushtkaran janyatvach mithyaityavgamyatai yatha kamladidoshat peetshankhgyanan ksyachit, atr cha jagatgyanasya sarvaishan srvada stvann doshosti*)-And it is held that that is false which is the result of a depraved cause, *e.g.* someone's cognition of a white conch-shell as yellow, through such a fault as the jaundice which depraves his eye-sight. But in the case in question-that of the world regarded as a reality, there is no such temporary or occasional depravation of the sense because all at all times cognize the world as a reality. Therefore the world is not an unreality.

Again he says: *Nasdutpado nrinshrigavt*-The production of that which does not already exist potentially is impossible like the horn of a man. *Upadananiymat*-Because there must of necessity be a material out of which a product is developed. Srvatr *srvada srvasanbhvat*-Because everything is not possible everywhere and always (which might be the case if materials could be dispensed with). The meaning is this : *Srvatr srvasmin Daiichi srvada srvasmin kalai srvanutpatairlokdrshanat*-In the world we see that everything is not possible everywhere and at all times. And *Shaktasy shakyekaran at*-Because anything possible must be produced from something competent to produce it.

In short, the Hindu philosopher's belief in the eternity of the world's substance arises from the fixed *article 'Ex nihilo nihil fit,'* nothing is produced out of anything. All the ancient philosophers of Greece-who are believed to have borrowed their theories from India-seem to have agreed upon this point. Lucretius starts with laying down the same principal. He says: "It things proceed from nothing, everything might spring from everything and nothing would require a seed. Men might arise first from sea, and fish and birds from earth, and flocks and herds break into being from sky; every kind of beast might be produced at random in cultivated places or deserts. The same fruits would not grow on the same trees but would be changed. All things would be able to produce all things."

6. *Samkhya* philosophy then starts with an original primordial *tattva* or eternally existing essence called *Prakriti*-a word means that which evolves or produces everything else. Some philosophers translate this *Prakriti* by nature. Certainly, nature is anything but a good equivalent for *Prakriti*, which donates something very different from matter or even germ of mere material substances. It is an intensely subtle original essence, wholly distinct from soul yet capable of evolving out of itself consciousness and mind as well as the whole visible world. In my opinion it is not even the name for anything which ever existed by itself. For Kapila himself in his work says: *Parnparyaipaikatr parinishthaiti sangyamatràm*

   In the manifestation of objects there must be a succession of causes without any end; and in Hindu logic the ruling idea is that you must suppose a point to exist where you should halt and *Prakriti* is only a halting point; therefore, it is in Kapila's words only a *sangyamatram, i.e.* merely a name given to the point in question, a mere sign to donate the cause which is the root which must be assumed rootless, merely to conform to the rule of Hindu logic.

7. Let us now see how Kapila defines this *Prakriti*. It is *Satvrajstamasan-Prakriti* is the state of equipoise of

Satv, Rajas, Tames goodness or passivity, passion, energy or activity and darkness or grossness. These three qualities passivity, activity and grossness-are not qualities in the ordinary sense. Qualities in the ordinary sense are attributes of *Prakriti*, they are rather the cords which when in a state of equipoise constitute *Prakriti*. On account of the disturbance of this state of equilibrium the whole world comes out. Kapila says: *Prakritairmhan mahatohankar ahankarat panchtanmatran i ubhyamindrayam ṭanmatraibhye sthoolbhootani Purush iti panchvinshtirgan* From *Prakriti* proceeds Mind *mehat*, from Mind self-consciousness, from self-consciousness the five subtle elements *Sthoolbhotani* and two sets of organs *Indriyas* external and internal, and from subtle elements gross elements *sthoolbhootani*. Thus *Prakriti* is the first basic primordial essence, and second principal evolved out of it is Mind, from Mind come out the third principal *Ahankara*, self-consciousness or individuality, from individuality come our five subtle elements and two sets of organs. These five subtle elements *are Shabd, Sparsh, Roop Ras Gandha*-sound, tangibility, form or colour, taste and smell or odour. The two sets of organs are external organs and internal organs. The external organs are again organs of sense and organs of action. The organs of sense are ear, skin, eye, nose, tongue; the organs of action are larynx, hand, foot, and the excretory and generative organs. These ten are external organs. The eleventh is the mind-the internal organ. From the five subtle elements are produced five gross elements-*Akash* (ether), *Vayu* (air), *Taijas* (fire or light), *Apas* (water), *Prithvi* (earth). The twenty-fifth is the *Purush*-the *Soul*, which is neither producer nor produced but eternal like *Prakriti*. It is quite distinct from the producing or produced elements and creation of the phenomenal world, though liable to be brought into connection with them.

8. The arguments which Kapila brings forward for the existence of soul as a separate entity, distinct from

Prakriti, are these; First, *Sanhatprarthatvat* that which combined and is therefore discreditable is finally for the sake of some other which is not discerptible. The second argument *Trigun adivipryat Soul* is something else than *Prakriti* because there is in Soul the reverse of the three qualities passivity, activity and grossness. The third argument is *Adhishthanach*-Soul is not material because of its superintendence over *Prakriti* (and a superintendent is an intelligent being while *Prakriti* is unintelligent). The fourth argument is *Bhoktribhavat*-Soul is not material because of its being the experiencer. It is the *Prakriti* that is experienced, the experiencer is soul.

What then is the nature of soul? Kapila answers: *Jadprkashayogat parkas* Since light does not pertain to the unintelligent, light is the essence of soul. The followers of the Vaisheshika system think that intelligence is only an attribute of soul; really it is without quality. It is essentially intelligent. If soul be unintelligent, it would not be a witness of its own comfort in profound and dreamless sleep. He does not agree with the *Vedantists* when they say that soul is one only for it is eternal, omnipresent, changeless, void of blemish; on the contrary, he says that from the fact [that] when one person is born another dies and a third one becomes old at the same time [it follows that] there is a multiplicity of souls. If soul were one only, when one is born all must be born. Both the *Vedantists* and the *Samkhya* are followers of the *Veda* and in the *Veda* there are passages like *Aikamaivadviteeyan brahm (chhandogyopanishad 6.2.1), naih nanasti kinchan (Vrihadan ykopnishad 4.4.19) mritio sa mritiomapnoti ye eh nanaiv pashyati (kathopnishad* 2.1.10)-Brahma is one without a second; there is nothing here diverse; death after death does he, the deluded man obtain who here sees as if it were a multiplicity. Kapila gives an ingenious interpretation to these passages. He says that his view of the multiplicity of souls is not opposed to the above passages of the Upanisads because those texts refer to the genus of all souls, *i.e.* to the fact that all souls are of the same nature. On the contrary he says in the *Puranas* we find passages to the effect that *Vamadeva* has been liberated, *Shuck* has been liberated. If soul were one, since the liberation of all would take place on the liberation of one the mention of diverse liberation's would be self-contradictory.

9. The soul is not considered by the *Samkhya* bound to matter. It is not bound, nor is it liberated. It is free. It has a delusive semblance of being bound. The nature of the soul is constant freedom and indifference to pleasure and pain alike.

10. These are the basic principals of the *Samkhya* philosophy. In short, according to its doctrines *Prakriti* and *Purush* are enough in themselves to and the idea of a creator is looked upon by the *Samkhya* as a mere redundant phantom of philosophy.

11. We may now enter into the details of this philosophy. In the first place let us ask Kapila what the motive is for the creation of the universe. He mentions two motives; they might have appeared satisfactory to him but to me his reply is not rational. He says that *Prakriti* created the universe for the emancipation of the soul which is really though not apparently emancipated or, secondly, for the removal of itself, *i.e.* for the sake of removing the actually real pain which consists of itself, as his commentator explains it. It the soul is essentially free and essentially light, there was no necessity for *Prakriti* to interfere with the soul's infinite bliss.

That soul is really though not apparently emancipated means that it is really emancipated but appears to be not so. Gandhi's interpretation of the phrase seems to be somewhat far-fetched, but he is apparently following some commentator. The more natural interpretation of the phrase should be: "Or we may say that *Prakriti* created the universe for the sake of itself, that is, for the sake of the removal of pain that really belongs to itself."As we have noted, in the *Samkhya* philosopher's eyes pain is a phenomenon belongsing to *Prakriti* rather than to *Purush*.

12. Let us examine the other stages of creation. I told you in the beginning that from *Prakriti* sprang the Great Mind. What is this Great Mind? Kapila says: It is intellect and judgement or ascertainment is its peculiar modification; and *Dharma, gyan, Vairagya, Aeshvarya i.e.* merit, knowledge, dispassion and supernatural power arise out of it when there is in it a superlative degree of the first if the three qualities, *i.e. Satv*, purity or passivity. But demerit, ignorance, non-dispassion and want of supernatural power arise out of it when there is in it a preponderance of the other two qualities. From the great principal-the Greet Mind, we were told, is produced *Ahankara i.e.* self-consciousness. It is what makes the Ego. It is the same as *Antakaran i.e.* the internal instrument. We were also told that the eleven organs and five subtle elements are produced from self-consciousness. But there is this distinction that the eleventh organ, the mind proceeds from self-consciousness in which the first quality *Satv*, purity or passivity-preponderates, while the other ten organs proceed from self-consciousness in which the second quality-activity or passion-predominates, and the five subtle elements proceed from self-consciousness in which the third quality-darkness or grossness-predominates. I have already enumerated the eleven organs. The popular opinion is that the organs are formed of gross elements. But the *Samkhya* doctrine is that is not so because the *Veda* does not support that view and we know that Kapila could not assume an attitude of direct opposition to the *Veda*s. There was another popular opinion about this mind-organ. It was that it is eternal, but Kapila says that none of the organs is eternal because the *Veda*s say so and because we see that they are destroyed. Further he says that mind is the leading organ while the other ten are kinds of powers. All these organs are mere instruments. As

a king even without himself taking an active part becomes a warrior simply by employing an army, so does the soul, although quiescent, through the different organs, become a seer, a speaker, a judge and the like, merely by reason of its proximity with these organs. There are some special properties belonging to the Great Intellect, self-consciousness and the mind. Attention or thought is the special property of the Intellect, conceit of personality is the property of self-consciousness, and decision and doubt of the mind, while the five airs-known as *Pran* etc.-are the common properties of all of them. The modifications of the organs are *Prman Vipreya, Vikalp, Nidra, Smriti,* evidence, chimera, sleep and memory. Some of them are painful, and others not painful. When these modifications cease to exist the soul comes to a state of self-quiet. The *Yoga* philosophy has the same doctrine. The very word *Yoga* means concentration and is defined as the suppression of the modifications of the thinking principal.

13. We will go still deeper into Kapila's philosophy. We have enumerated in the beginning the 25 principals commencing with *Prakriti* and ending with *Purusa*. *Prakriti* as *Prakriti* in a state of equilibrium is unable to produce anything. It is only when equilibrium is disturbed that the creation follows. *Purush*-the soul-itself is neither the producer nor the produced. Whence is the human body created according to this philosophy? Kapila says that out of the remaining twenty-three principals a pair of bodies *sthool shreer* and *Sooksham shreer* gross body and subtle body originates. In fact the twenty-three principals act as the seed. out of which the body is produced and the fact that the soul becomes conditioned by the 23 principals is the cause of its going from one body to another in fact the cause of all mundane existence, and this mundane existence continues for each soul so long as it does not discriminate the difference between soul and *Prakriti*. It should be noted how-ever that according to Kapila's theory the soul is

not really fettered by matter, it only has a wrong impression that it is fettered. Really it is quite free. Only it does not realize this fact so long as it is in mundane existence. We come again to the pair of bodies-the gross body and the subtle body. The gross body usually though not always arises from father and mother, while the subtle body is a creation out of the principals. Pleasure and pain belong to the subtle

Body, not to the gross body. In the beginning of the creation there was but one subtle body which consisted of the collection of seventeen elements-eleven organs, five subtle elements and the *Buddhi*, *i.e.* the great intellect the understanding. But through the diversity of actions later on the one subtle body became differentiated into many. The subtle body does not exist independently, It has its tabernacle-the gross body for residing therein. As a shadow or a picture does not stand without a support, so the subtle body at death leaves one gross body and passes into another. It cannot in fact exist independently because its essence is *Satvprakash* pure light and all luminous ether is seen only as associated with earthy substance. The gross body is a composition of the five gross elements.

14. What aims then are accomplished by the subtle body transmigrating from one gross body to another? Kapila says *Gyananmukti*. From knowledge (acquired through mundane existence) comes the liberation, *i.e.* the discrimination between soul and non-soul. Bondage is also one of the aims of this transmigration but it arises on account of misconception. Kapila altogether discards the theory of the efficacy of works as a means of salvation. To him only knowledge is the sole means of libertaion. Even meditation is not the direct cause of liberation, though it is useful as secondary cause, for it removes desire, which really hinders knowledge. So it is worth practicing, which can be done by stopping all modifications of the Mind. This is done by *dharn asan* and *Svkarm* restrain, posture and the fulfillment

of duties. By restrain I mean the restrain of breath by means of expulsion and retention under certain rules. By posture is meant the peculiar position in sitting gives pleasure, and by the fulfillment of the duties is meant by the performance of actions prescribed for one's religious order. This meditation can be acquired only through *Vairagya* and *abhyasa* dispassion and constant practice. Through meditation knowledge is acquired. But if misconception interferes, bondage will be the result. What is this misconception? It is fivefold *Avidya Asmat*, *Rag, Dvaish* and *Abhinivaish* ignorance, egoism, attachment, aversion and fear of dissolution. Why should this misconception play its part at all? Simply because the powers called *Tushti* and *Siddhi* are impeded and hence arises the disability which cause misconception. Much can be said with reference to these powers of *Tushti* and *Siddhi*. But our time will not permit us to go into any details. We shall come to some of them when we shall talk on the *Yoga* philosophy.

15. There is however one point to which, I should draw your attention. I mean the nature of the *Samkhya Mukti* the liberation of the soul. His theory is not, as misunderstood by Western orientalists, the theory of absorption. The soul on liberation does not merge into the Universal Spirit or into the Absolute, for in his system there is no such thing as the Supreme Spirit or the Absolute. Not only does he not propound such a theory as the final object but on the contrary he refutes it. He thinks that by merging into the primordial original essence, the *Prakriti*, the souls will have to rise again and pass through different mundane existence. It is only when the right discrimination of soul and non-soul takes place that there will be the final emancipation of the soul.

16. There is another point to which I should like to draw your attention. The *Samkhya* philosophy in a large measure supports the nature working under fixed laws without any interference on the part of an extra-cosmic being.

17. But of all his theories, one that has struck me to be the most liberal is the universal salvation theory. He does not restrict the liberation only to the few followers of his philosophy but to others also.
18. So far we have tried to understand the meaning of Kapila's theory. Let us now see if it is consistent and appeals to our reason. In the first place, he says that *Prakriti* was in the beginning in a state of equilibrium. The three qualities, passivity, activity and grossness, were balanced. What then caused a disturbance in this state of equilibrium? Without external-causes, *Prakriti* cannot be disturbed. *Pursha* the soul is action-less, changeless, without any qualities or attributes.

Secondly, the Great Mind and self-consciousness are considered by Kapila to be different form each other. According to him one is the product of the other. And both of them are the outcome of *Prakriti*, which is really material. Now the Great Mind or *Buddhi* or intellect is nothing but a phase of consciousness. Self-consciousness-'I am happy ','I am unhappy'- is only a particular instance illustrating that phase and both of them imply knowledge and are but the characters of the soul but can never be the products of primordial material essence.

With regard to subtle elements Kapila says that gross elements are produced from these subtle elements; *e.g.*, from odour comes out earth, from taste water, from colour fire, from touch wind and from sound ether. If he means that the gross elements, which we see outside the human or any other gross organic body, are the products of these subtle elements, there is no reason to support it. The external elements we see are as eternal as anything else.

The Samkhya Aphorisms of Kapila

a. Salutation to the illustrious sage, Kapila!

b. Well, the great sage, Kapila, desirous of raising the world [from the Slough of Despond in which he found it sunk], perceiving that the knowledge of the *excellence* of any fruit, through the desire [which this excites] for the fruit, is a cause of people's betaking themselves to the means [adapted to the attainment of the fruit], declares [as follows] the excellence of the fruit [which he would urge our striving to obtain]:

Aph. 1.* Well, the complete cessation of pain [which is] of three kinds is the complete end of man.

a. The word 'well' serves as a benediction; [the particle *atha* being regarded as an auspicious one].

b. By saying that the complete cessation of pain, which is of three kinds,—*viz.*, (1) due to one's self (*adhyatmika*), (2) due to products of the elements (*adhibhautika*), and (3) due to supernatural causes (*adhidaivika*),—is the *complete* end of man, he means to say that it is the *chief* end of man, among the four human aims, [*viz.*, merit, wealth, pleasure, and *liberation* (see *Sahitya-darpaIa*, § 2)]; because the three are transitory, whereas liberation is *not* transitory: such is the state of the case.

c. But then, let it be that the above-mentionend cessation [of all the three kinds of pain] is the complete end of man; still, what reason is there for betaking one's self to a doctrinal system which is the cause of a knowledge of the truth, in the shape of the knowledge of the difference between Nature and Soul, when there are *easy* remedies for bodily pains, *viz.*, drugs, &c., and remedies for mental pains, *viz.*, beautiful women and delicate food, &c., and remedies for pains due to products of the elements, *viz.*, the residing in impregnable localities, &c., as is enjoined in the institutes of polity, and remedies for pains due to supernatural causes, *viz.*, gems [such as possess marvellous prophylactic properties], and spells, and herbs of mighty power, &c.; and when [on the other hand], since it is hard to get one to grapple with that very difficult knowledge of truth which can be perfected only by the toil of many successive births, it must be still *more* hard to get one to betake himself to the doctrinal system [which treats of the knowledge in question]? Therefore [*i.e.*, seeing that this may be asked] he declares [as follows]:

Aph. 2.* The effectuation of this [complete cessation of pain] is not [to be expected] by means of the visible [such as wealth, &c.]; for we see [on the loss of wealth, &c.,] the restoration [of the misery and evil,] after [its temporary] cessation.

a. 'The visible,' in the shape of the drugs, &c., above-mentioned [§ 1. *c.*].

b. 'The effectuation of this,' *i.e.*, the effectuation of the complete cessation of pain.

c. Why is it not [to be thus effected]? Because, after the cessation (the cessation of pain is understood), we see its restoration, the springing up again of pain in general, [from whichever of its three sources (§ 1. *b.*)].

d. The state of the matter is this: not by the expedients above-mentioned is there such a removal of pain, that no pain arises thereafter; for, when, by this or that expedient, this or that pain has been destroyed, we see other pains springing up. Therefore, though it be *not* easy [§ 1. *c.*], the knowledge of truth [as a complete remedy] *is* to be desired.

e. But then, grant that *future* pain is not debarred by drugs, &c., [employed to remove present pain], still, by again and again obviating it [as often as it presents itself], there may be the cessation of *future* pain, also. This doubt he states [as follows]:

Aph. 3.* [Let us consider the doubt] that the soul's desire [the cessation of pain, may result] from exertions for the obviation [of pain], as is the case with the obviation of daily hunger

a. When pain shall arise [let us suppose one to argue] then it is to be obviated; and thus there is the soul's desire, the cessation of pain; just as one should eat, when there is hunger; and thus there is the soul's desire of the eater, *viz.*, the cessation of hunger. In regard to this [doubt] he states the recognized decision:

Aph. 4.* This [method of palliatives (§ 3)] is to be rejected by those who are versed in evidence; because it is not everywhere possible [to employ it at all], and because, even if this *were* possible, there would be an impossibility as regards [ensuring] the perfect fitness [of the agents employed].

a. For there are not physicians, &c., in every place and at all times; and [to rely on physicians, &c., would not be advisable], even if there were the possibility,—*i.e.*, even if these *were* [always at hand], since physicians are not *perfect* [in their art];—for pain cannot with certainty be got rid of by means of physicians, &c., with their drugs, &c. Moreover, when corporeal pain has departed, there may still be that which is *mental*, &c.; so that there is not [under such circumstances], in every respect, liberation from pain. For these reasons, *such* a soul's aim [as that which contents itself with temporary palliatives] is to be rejected by those who are versed in evidence, [*i.e.*, who are acquainted with authoritative treatises].

b. He mentions another proof [of his assertion]:

Aph. 5.* Also [an inferior method ought not to be adopted] because of the preeminence of Liberation [as proved] by the text [of Scripture declaratory] of its preeminence above all else.

a. One ought not to endeavour after the removal of this or that pain by these and those expedients [§ 1. *c.*]; since Liberation (*moksha*), by being eternal, is transcendent as a remover of all pains. Moreover, one ought to endeavour only after the knowledge of truth, which is the means thereof [*i.e.*, of Liberation]; because the Scripture tells its preeminence above all [other objects of endeavour], in the text: 'There is nothing beyond the gaining of Soul, [with the utter exclusion of pain].'

b. But then [it may be suggested], when you say *liberation*, we understand you to mean from *bondage*. And is that bondage essential? Or is it adventitious? In the former case, it is incapable of destruction; if it come under the latter head, it will perish of itself, [like any other adventitious and, therefore, transitory thing]. What have we to do with your 'knowledge of truth,' then? To this he replies [as follows]:

Aph. 6.* And there is no difference between the two.

a. There is no difference in the applicability of liberation, on either of the suppositions, that the bondage is essential, and that it is adventitious, [supposing it were either (see § 19. *b.*)]. That is to say, we can tell both how the bondage takes place, and how the liberation takes place.

b. Now, with the view of demonstrating [the real nature of] Bondage and Liberation, he declares, exclusively, in the first place, the objections to Bondage's being *essential* [§ 5. *b*.]:

Aph. 7.* There would be no rule in the enjoining of means for the liberation of one bound *essentially*.

a. Since Liberation has been stated [§ 1] to result from the complete cessation of pain, [it follows that] Bondage is the junction of pain; and this is not *essential* in man. For, if that were the case, then there would be no rule, *i.e.*, no fitness, in the Scriptural or legal injunction of means for liberation: such is what must be supplied, [to complete the aphorism]. Because, to explain our meaning [by an illustration], *fire* cannot be liberated from its *heat*, which is eseential to it; since that which is *essential* exists as long as the substance exists.

b. And it has been declared in the Divine Song [the *Ishwara-gita*,]: 'If the soul were essentially foul, or impure, or changeable, then its liberation could not take place even through hundreds of successive births.'

c. [Since some one may be disposed to say] '*Grant* that there is no fitness [in the Scriptural and legal injunctions, (§ 7. *a*.)], what have we to do with *that*?' Therefore he declares [as follows]:

Aph. 8.* Since an essential nature is imperishable, unauthoritativness, betokened by impracticableness, [would be chargeable against the Scripture, if pain were essential to humanity].

a. That is to say: since the essential nature of anything is imperishable, *i.e.*, endures as long as the thing itself, it would follow [on the supposition that pain is essential to humanity], that, since Liberation is *impossible*, the Scripture which enjoins the means for its attainment is a false authority, inasmuch as it is impracticable [in its injunctions. And this is out of the question; Scripture being assumed, here, as in all the others of the six systems, to be an exact measure of truth].

b. But then [some one may say], let it *be* an injunction [to use means for the attainment of an unattainable object], on the mere strength of Scripture; [and, since Scripture is an unquestionable authority, we may be excused from asking or answering the question, *why* the injunction is given]. To this he replies [as follows]:

Aph. 9.* There is no rule, where something impossible is enjoined: though it *be* enjoined, it is no injunction.

a. There can be no fitness, or propriety, in an injunction with a view to an impossible fruit; seeing that, though something be enjoined, or ordered [to be effected] by means that are impracticable, this is no injunction at all, but only the *semblance* of an injunction; because it stands to reason, that not even the *Veda* can make one see sense in an absurdity: such is the meaning.

b. Here he comes upon a doubt:

Aph. 10.* If [some one says] as in the case of white cloth, or of a seed, [something essential may be not irremovable, then he will find his answer in the next aphorism].

a. But then [the doubter is supposed to argue], the destruction even of what is essential [in spite of what is stated under *s* seen; as, for example, the essential whiteness of white cloth is removed by dyeing, and the essential power of germination in a seed is removed by fire. Therefore, according to the analogy of the white cloth and the seed, it is possible that there should be the removal of the bondage of the soul, even though it *were* essential. So, too, there may be [without any impropriety] the enjoinment of the means thereof. Well, *if* [any one argues thus], such is the meaning [of the aphorism, to which he proceeds to reply].

b. He declares [the real state of the case, with reference to the doubt just raised]:

Aph. 11.* Since both perceptibleness and [subsequent] non-perceptibleness may belong to some power [which is indestructible], it is not something *impracticable* that is enjoined, [when one is directed to render some indestructible* power imperceptible].

a. In regard even to the two examples above-mentioned [§ 10], people do not give an injunction for [the positive destruction of]** something essential, which is indestructible

[§ 8]. Why [do we say this]? Because, in these two instances of the perceptibleness and non-perceptibleness of a power [the powers, namely, of appearing white and of germinating (see § 10. *a*.)], there are merely the manifestation and [afterwards] the *hiding* of the whiteness, &c., but not the *removal* of the whiteness, or of the power of germination; because, that is to say, the whiteness of the dyed cloth and the germinating power of the roasted seed can again be brought out by the processes of the bleacher, &c., [in the case of the dyed cloth], and by the will of the *Yogi*, [the possessor of supernatural powers, in the case of the roasted seed], &c.

b. Having thus disproved the notion that bondage is *essential* [to man], wishing to disprove also the notion that it is the result of some [adherent] *cause*, be rejects the [various supposable] causes, *viz.*, Time, &c.:

Aph. 12.* Not from connexion with *time* [does bondage befall the soul]; because this, all-pervading and eternal, is [eternally] associated with *all*, [and not with those alone who are in bondage].

a. The bondage of man is not caused by *time*; because [if that were the case,] there could be no such separation as that of the *liberated* and *unliberated*; because time, which applies to everything, and is eternal, is at all times associated with all men, [and must, therefore, bring *all* into bondage, if any].

Aph. 13.* Nor [does bondage arise] from connexion with *place*, either, for the same [reason]

a. That is to say: bondage does not arise from connexion with *place*. Why? 'For the same reason,' *i.e.*, for that stated in the preceding aphorism, *viz.*, that, since it [*viz.*, place] is connected with *all* men, whether liberated or not liberated, bondage would [in *that* case] befall the *liberated*, also.

Aph. 14.* Nor [does the bondage of the soul arise] from its being conditioned [by its standing among circumstances that clog it by limiting it]; because *that* is the fact in regard to [not the soul, but] the *body*.

a. By 'condition' we mean the being in the shape of a sort of association. The bondage [of the soul] does not arise from *that*; because *that* is the property of the *body* [and not of the soul]; because, that is to say, bondage might befall even the liberated [which is impossible], if that which is the fact in regard to another could occasion the bondage of one quite different.

b. But then [some one might say], *let* this conditioned state belong to the soul. On this point [to prevent mistakes], he declares:

Aph. 15.* Because this soul is [unassociated with any conditions or circumstances that could serve as its bonds, it is] absolute.

a. The word *iti* here shows that it [*i.e.*, the assertion conveyed in the aphorism] is a *reason*; the construction with the preceding aphorism being this, that, *since* the soul is unassociated, it belongs only to the body to be conditioned.

Aph. 16.* Nor [does the bondage of soul arise] from any work; because [works are] the property of another [*viz.*, the mind], and because it [the bondage] would be eternal, [if the case were as you imagine].

a. That is to say: moreover, the bondage of the soul does not arise from any work, whether enjoined or forbidden; because works are the property of another, *i.e.*, not the property of the soul [but of the mind]. And, if, through a property of another, the bondage of one quite distinct could take place, then bondage might befall even the liberated, [through some acts of some one else].

b. But then [some one may say], this objection does not apply, if we hold that bondage may arise from the acts of the *associate* [*viz.*, the mental organ]: so, with allusion to this, he states another reason, 'and because it would be eternal,' *i.e.*, because bondage, in the shape of connexion with pain, would occur [where it does not,] even in such cases as the universal dissolutions [of the phenomenal universe, including the mental organ, but *not* the soul].

c. But then [some one may say], if that be the case, then let the bondage, too, in the shape of connexion with pain, belong [not to the *soul*, but] to the *mind* alone, in accordance with the principle that it have the same locus as the works [to which it is due]; and, since it is an established point that pain is an affection of the *mind*, why is bondage [*i.e.*, connexion with pain] assumed of the *soul*, also? With reference to this doubt, he declares [as follows]:

Aph. 17.* If it were the property of any other, then there could not be diverse experience.

a. If bondage, in the shape of connexion with pain, were the property of another, *i.e.*, a property of the *mind*, there could be no such thing as diverse experience; there could be no such different experience as one man's experiencing pain, and another man's not: [for, it must be remembered, it is not in point of *mind*, but of *soul*, that men are held, by Kapila, to be numerically different]. Therefore, it must be admitted that pain is connected with the soul, also. And this [pain that belongs to the soul] is in the shape merely of a *reflexion* of the pain [that attaches to its attendant organism]; and this reflexion is of its *own* attendant [organism] only; so that there is no undue result [deducible from our theory].

b. He rejects also the notion that Nature (*prak[iti*) is *directly* the cause of bondage:

Aph. 18.* If [you say that the soul's bondage arises] from Nature, as its cause, [then I say] 'no;' [because] that, also, is a dependent thing.

a. But then [some one may say], let bondage result from *Nature*, as its cause. If you say so, I say 'no;' because that, also, *i.e.*, Nature, also, is dependent on the *conjunction* which is to be mentioned in the next aphorism; because, if it [Nature] were to occasion bondage, even *without* that [conjunction which is next to be mentioned], then bondage would occur even in such cases as the universal dissolution, [when soul is altogether disconnected from the phenomenal].

b. If the reading [in the aphorism] be *nibandhana* [in the 1st case, and not in the 5th], then the construction will be as follows: 'If [you say that] the bondage is caused by Nature,' &c.

c. Therefore, since Nature can be the cause of bondage, only as depending on something else [*i.e.*, on the conjunction to be mentioned in the next aphorism], through this very sort of conjunction [it follows that] the bondage is *reflexional*, like the heat of water due to the conjunction of fire; [water being held to be essentially cold, and to *seem* hot only while the heat continues in conjunction with it].

d. He establishes his own tenet, while engaged on this point, in the very middle [of his criticisms on erroneous notions in regard to the matter; for there are more to come]:

Aph. 19.* [But] not without the conjunction thereof [*i.e.*, of Nature] is there the connexion of that [*i.e.*, of pain] with that [*viz.*, the soul,] which is ever essentially a pure and free intelligence.

a. Therefore, without the conjunction thereof, *i.e.*, without the conjunction of Nature, there is not, to the soul, any connexion with that, *i.e.*, any connexion with bondage; but, moreover, just through that [connexion with Nature] does bondage take place.

b. In order to suggest the fact that the bondage [of the soul] is reflexional [and not inherent in it, either essentially or adventitiously], he makes use of the indirect expression with a double negative, ['not without']. For, if bondage were produced by the conjunction [of the soul] with Nature, as colour is produced by heating [in the case of a jar of black clay, which becomes red in the baking], then, just like that, it would continue even after disjunction therefrom; [as the red colour remains in the jar, after the fire of the brick-kiln has been extinguished, whereas the red colour occasioned in a crystal vase by a China-rose, while it occurs *not without* the China-rose, ceases, on the removal thereof]. Hence, as bondage ceases, on the disjunction [of the soul] from Nature, the bondage is merely reflexional, and neither essential [§ 5. *b*.] nor adventitious [§ 11. *b*.].

c. In order that there may not be such an error as thot of the Vais*h*eshikas, *viz.*, [the opinion that there is] an absolutely real conjunction [of the soul] with pain, he says 'which is ever,' &c. [§ 19]. That is to say: as the connexion of colour with essentially pure crystal does not take place without the conjunction of the China-rose [the hue of which, seen athwart the crystal, seems to belong to the crystal], just so the connexion of *pain* with the soul, ever essentially pure, &c., could not take place without the conjunction of some accidental associate; that is to say, pain, &c., cannot arise spontaneously, [any more than a red colour can arise *spontaneously* in the crystal which is essentially pure].

d. This has been declared, in the *Saura*, as follows 'As the pure crystal is regarded, by people, as red, in consequence of the proximity of something [as a China-rose] that lends its colour, in like manner the supreme soul [is regarded as being affected by pain].'

e. In that [aphorism, 19], the perpetual purity means the being ever devoid of merit and demerit; the perpetual intelligence means the consisting of uninterrupted thought; and the perpetual liberatedness means the being ever dissociated from *real* pain: that is to say, the connexion with pain in the shape of a *reflexion* is not a real bondage, [any more than the reflexion of the China-rose is a real stain in the crystal].*

f. And so the maker of the aphorism means, that the cause of its bondage is just a particular *conjunction* [§ 19. *c*.]. And now enough as to that point.

g. Now he rejects [§ 18. *d*.] certain causes of [the soul's] bondage, preferred by others:

Aph. 20.* Not from Ignorance, too, [does the soul's bondage arise]; because that which is not a reality is not adapted to binding.

a. The word 'too' is used with reference to the previously mentioned 'Time,' &c., [§ 12, which had been rejected, as causes of the bondage, antecedently to the statement, in § 19, of the received cause].

b. Neither, too, does [the soul's] union with bondage result directly from 'Ignorance,' as is the opinion of those who assert non-duality [or the existence of no reality save one (see *Vedanta-sara*, § 20. *b*.)]; because, since their 'Ignorance' is not a real thing, it is not fit to bind; because, that is to say, the binding of any one with a rope merely *dreamt* of was never witnessed.

c. But, if 'Ignorance' *be* a reality [as some assert], then he declares [as follows]:

Aph. 21.* If it ['Ignorance'] *be* [asserted, by you, to be] a reality, then there is an abandonment of the [Vedant*i*c] tenet, [by you who profess to follow the Vedanta].

a. That is to say: and, if you agree that 'Ignorance' *is* a reality, then you abandon your own implied dogma* [see Nyaya Aphorisms I., § 31] of the unreality of Ignorance;' [and so you stultify yourself].

b. He states another objection:

Aph. 22.* And [if you assume 'Ignorance' to be a reality, then] there would be a *duality*, through [there being] something of a different kind [from soul; which you asserters of *non-duality* cannot contemplate allowing].

a. That is to say: if 'Ignorance' is real and without a beginning, then it is eternal, and coordinate with Soul: if [therefore] it be *not* soul, then there is a duality, through [there being] something of a different kind [from soul; and this the Vedant*i*s cannot intend to establish]; because these followers of the *Vedanta*, asserting *non-duality*, hold that there is neither a duality through there being something of the same kind [with soul], nor through there being something of a different kind.

Aph. 23.* If [the Vedant*i* alleges, regarding 'Ignorance,' that] it is in the shape of both these opposites, [then we shall say 'no,' for the reason to be assigned in the next aphorism].

a. The meaning is: if [the Vedant*i* says that] 'Ignorance' is not *real*,—else there would be a duality through [there being] something of a different kind [from soul, which a follower of the Vedanta cannot allow],—and, moreover, it is not *unreal*, because we experience its effects; but it is in the shape of something at once real and unreal, [like Plato's B*i* êav ìt D*i*: (see *Vedanta-sara*, § 21)].

Aph. 24.* [To the suggestion that 'Ignorance' is at once real and unreal we say] 'no;' because no such thing is known [as is at once real and unreal.]

a. That is to say: it is not right to say that 'Ignorance' is at once real and unreal. The reason of this he states in the words 'because no such thing,' &c.; because any such thing as is at once real and unreal is not known. For, in the case of a dispute, it is neccssary that there should be an *example* of the thing [*i.e.* (see Nyaya Aphorisms, I., § 25), a case in which all parties are agreed that the property in dispute is really present]; and, as regards *your* opinion, such is not to be found; [for, where is there anything in regard to which both parties are agreed that it is at once real and unreal, as they are agreed that fire is to be met with on the culinary hearth?]: such is the import.

b. Again he ponders a doubt:

Aph. 25.* [Possibly the Vedant*i* may remonstrate] '*We* are not asserters of any Six Categories, like the *Vaisheshikas* and others.'

a. 'We are not asserters of a definite set of categories [like the *Vaisheshikas*, who arrange all things under six heads, and the *Naiyayikas*, who arrange them under sixteen]. Therefore, we hold that there *is* such a thing, unknown though it be [to people in general], as 'Ignorance' which is at once real and unreal, or [if you prefer it], which differs at once from the real and the unreal [see *Vedanta-sara*, § 21]; because this is established by proofs,' [Scriptural or otherwise, which are satisfactory to *us*, although they may not comply with all the technical requisitions of Gotama's scheme of argumentative exposition (see Nyaya Aphorisms, I., § 35)].

b. By the expression [in the aphorism] 'and others' are meant the *Naiyayikas*; for the *Naiyayika* is an asserter of sixteen categories [see Nyaya Aphorisms, I., § 1].

c. He confutes [this pretence of evading the objection, by disallowing the categories of the Nyaya]:

Aph. 26.* Even although this be not compulsory [that the categories be six, or sixteen], there is no acceptance of the inconsistent; else we come to the level of children, and madmen, and the like.

a. Let there be [accepted] no system of categories [such as that of the *Vaisheshika,* § 25]; still, since *being* and *not-being* are contradictory, it is impossible for disciples to admit, merely on Your Worship's assertion, a thing at once real and unreal, which is inconsistent, contrary to all fitness: otherwise, we might as well accept also the self-contradictory assertions of children and the like: such is the meaning.

b. Certain heretics [deniers of the authority of the Vedas] assert that there exist external objects of momentary duration [individuaily; each being, however, replaced by its facsimile the next instant, so that the uninterrupted series of productions becomes something equivalent to continuous duration], and that by the influence of these the bondage of the soul [is occasioned]. This he objects to, [as follows]:

Aph. 27.* [The bondage] thereof moreover, is not caused by any influence of objects from all eternity.

a. 'Thereof,' *i.e.*, of the soul. An eternal influence of objects, an influence of objects the effect of which, in the shape of a continued stream, has had no commencement,—not by *this*, either, is it possible that the bondage [of the soul] has been occasioned: such is the meaning.

b. He states the reason of this [impossibility]:

Aph. 28.* Also [in my opinion, as well as in yours, apparently], between the external and the internal there is not the relation of influenced and influencer; because there is a local separation; as there is between him that stays at Srughna and him that stays at Pamaliputra.

a. In the opinion of these [persons whose theory we are at present objecting to], the soul is circumscibed, residing entirely within the body; and that which is thus *within* cannot stand in the relation of the influenced and the influencer, as regards an *external* object. Why? Because they are separated in regard to place; like two persons the one of whom remains in Srughna and the other in Pamaliputra: such is the meaning. Because the affection which we call 'influence' (*vasana*) is seen only when there is conjunction, such as that of madder and the cloth [to which it gives its colour], or that of flowers and the flower-basket [to which they impart their odour.]

b. By the word 'also' the absence or conjunction [between the soul and objects (see § 10)], &c., which he himself holds, is connected [with the matter of the present aphorism].

c. Srughna and Pamaliputra [Palibothra, or Patna] are two several places far apart.

d. But then [these heretics may reply], 'The influence of objects [on the soul] may be asserted, because there is a contact with the object; inasmuch as the soul, according to *us*, goes to the place of the object, just as the senses, according to Your Worship.' Therefore he declares [as follows]:

Aph. 29.* [It is impossible that the soul's bondage should arise] from an influence received in the same place [where the object is; because, in that case], there would be no distinction between the two, [the bond and the free].

a. To complete the sense, we must supply as follows: 'It is impossible that the bondage should arise from an influence received in one and the same place with the object.' Why? Because there would be no distinction between the two, the soul bound and the soul free; because bondage would [in that case] befall the liberated soul, also; [the free soul, according to this hypothesis, being just as likely to come across objects as any other]: such is the meaning.

b. Here he ponders a doubt:

Aph. 30.* If [the heretic, wishing to save his theory suggests that a difference between the two cases (see § 29) *does* exist] in virtue of the *unseen*, [*i.e.*, of merit and demerit, then he will find his answer in the next aphorism].

a. That is to say, [the heretic may argue]: 'But then, granting that they [the free soul and the bound] are alike in respect of their coming into contact with objects, when they become conjoined with them in one and the same locality; yet the *reception of the influence* may result merely from the force of the *unseen*, [*i.e.*, from the merit and demerit of this or that soul; the soul that is liberated alike from merit and demerit being able to encounter, with impunity, the object that would enchain one differently circumstanced]': if [*this* be urged, then we look forward].

b. This he disputes, [as follows]:

Aph. 31.* They cannot stand in the relation of deserver and bestower, since the two do not belong to one and the same time.

a. Since, in thy opinion, the agent and the patient are distinct, and do not belong to the same time [believing, as thou heretically dost, not only that *objects* (see § 26. *b*.) momentarily perish and are replaced, but that the duration of *souls*, also, is of a like description], there is positively no such relation [between the soul at one time and its successor at another] as that of deserver and bestower [or transmitter of its merits or demerits]; because it is impossible that there should be an influence of objects [§ 27] taking effect on a patient [say, the soul of to-day], occasioned by the 'unseen' [merit or demerit] belonging to an agent [say, the soul of yesterday, which, on the hypothesis in question, is a numerically different individual]: such is the meaning.

b. He ponders a doubt:

Aph. 32.* If [the heretic suggests that] the case is like that of the ceremonies in regard to a son, [then he will find his reply by looking forward].

a. But then [the heretic, admitting the principle that the merit or demerit of an act belongs entirely to the agent, may urge that], as the son is benefited by ceremonies in regard to a son, such as that [ceremony (see Colebrooke's 'Hindu Law,' Vol. III.) celebrated] in anticipation of conception, which [no doubt] belongs to the *father* [who performs the ceremonies, to propitiatc the gods], in like manner there may be an influence of objects on the experiencer [say, the soul of to-day], through the 'unseen' [merit or demerit] that belongs even to a different subject [say, the soul of yesterday]: such is the meaning [of the heretic].

b. He refutes this, by showing that the illustration is not a fact:

Aph. 33.* [Your illustration proves nothing;] for, in that case, there is no one permanent soul which could be consecrated by the ceremonies in anticipation of conception, &c.

a. 'In that case,' *i.e.*, on thy theory, too, the benefit of the son, by [means of the performance of] the ceremonies in anticipation of conception, &c., could not take place; 'for,' *i.e.*, because, on that theory, there is not one [self-identical] soul, continuing from the [time of] conception to birth, which could be consecrated [by the ceremonies in question], so as to be a fit subject for the duties that pertain to the time subsequent to birth [such as the investiture with the sacred thread, for which the young Brahman would not be a fit subject, if the ceremonies in anticipation of his conception had been omitted]: and thus your illustration is not a real one, [on your *own* theory: it is not a thing that you can assert as a fact].

b. And, according to *my* theory, also, your illustration is not a fact; seeing that it is possible that the benefit to the son should arise from the 'unseen' [merit] deposited in the son by means of the ceremony regarding the son: for it is an implied tenet [of my school], that it [the soul] is permanent [in its self-identity]; and there is the injunction [of Manu, (Ch. II., v. 26), with regard to the ceremonies in question, which proceeds on the same grounds].

c. Some other heretic may encounter us, on the strength of [the argument here next stated, *viz.*,] 'But then, since *bondage*, also, [like everything else] is momentary, let this bondage have nothing determinate for its cause, or *nothing at all* for its cause,' [which view of matters is propounded in the next aphorism]:

Aph. 34.* Since there is no such thing as a permanent result [on the heretical view], the momentariness [of bondage, also, is to be admitted].

a. 'Of bondage': this must be supplied, [to complete the aphorism].

b. And thus the point relied on is, that it [*i.e.*, bondage] have no cause at all. And so this is the application [of the argument, *viz.*]:

(1) Bondage, &c., is momentary;

(2) Because it exists,

(3) [Everything that exists is momentary,] as the apex of the lamp-flame, or the like.

c. And [continues the heretic,] this [reason, *viz.*, 'existence'] does not extend *unduly* [as you may object,] to the case of a jar, or the like; because *that*, also [in my opinion], is like the subject in dispute; [in being momentary]. This [in fact] is precisely what is asserted in the expression, 'since there is *no such thing* as a permanent result' [§ 34].

d. He objects [to this heretical view]:

Aph. 35.* No, [things are *not* momentary in their duration]; for the absurdity of this is proved by *recognition*.

a. That is to say: nothing is momentary; because the absurdity of its being momentary follows from the opposite argument [to that under § 34. *b.*]. taken from such facts of recognition as, 'what I saw, that same do I touch,' [an argument which may be stated as follows], *viz.*:

(1) Bondage, &c., is permanent;

(2) Because it exists,

(3) [Everything that exists is permanent,] as a jar, or the like.

Aph. 36.* And [things are not momentary;] because this is contradicted by Scripture and reasoning.

a. That is to say: nothing is momentary; because the general principle, that the whole world, consisting of effects and causes, is momentary, is contradicted by such texts as this, *viz.*, '[All] this, O ingenuous one, was antecedently existing,' and by such Scriptural and other arguments as this, *viz.*, 'How should what exists proceed from the non-existent?'

Aph. 37.* And [we reject the argument of this heretic;] because his instance is not a fact.

a. That is to say: the general principle of the momentariness [of all things] is denied; because this momentary churacter does not [in fact] belong to the apex of the lamp-flame, &c., the instance [on which thou, heretic, dost ground thy generalization, (§ 34. *b.*)]. Moreover, thou quite errest in regard to momentariness, in that instance, from not taking account of the minute and numerous instants [really included in a duration which seems to thee momentary]: such is the import.

b. Moreover, if the momentary duration, &c., [of things] be asserted, then there can be no such thing as the relation of cause and effect, in the case of the earth and the jar, and the like. And you must not say that there *is* no such thing as that [relation of cause and effect]; because it is proved to be a reality by the fact that, otherwise, there would be no such thing as the efforts of him who desires an effect; [and who, therefore, sets in operation the causes adapted to its production]. With reference to this, he declares [as follows]:

Aph. 38.* It is not between two things coming simultaneously into existence, that the relation of cause and effect exists.

a. Let us ask, does the relation of product and [material] cause exist between the earth and the jar, as *simultaneously* coming into [their supposed momentary] existence, or as successive? Not the first; because there is nothing to lead to such an inference, and because we should not [in that case] find the man, who wants a jar, operating with earth, &c., [with a view to the jar's *subsequent* production]. Nelther is it the last; in regard to which he declares [as follows]:

Aph. 39.* Because, when the antecedent departs the consequent is unfit [to arise, and survive it].

a. The relation of cause and effect is, further, inconsistent with the theory of the momentary duration of things; because, at the time when the antecedent, *i.e.*, the cause, departs, the consequent, *i.e.*, the product, is 'unfit,' *i.e.*, is not competent to arise; because, that is to say, a product is cognized only by its inhering in [and being substantially identical with, however formally different from,] its substantial cause, [and is incapable, therefore, of surviving it].

b. With reference to this same [topic, *viz.*, the] substantial cause, he mentions another [the converse] objection [to the theory of the momentary duration of things]:

Aph. 40.* Moreover, not [on the theory of the momentary duration of things can there be such a relation as that of cause and effect]; because, while the one [the antecedent] exists, the other [the consequent] is incompatible, because the two keep always asunder.

a. To complete [the aphorism], we must say, 'moreover, [on the theory objected to], there can be no such relation as that of cause and effect; because, at the time when the antecedent exists, the consequent cannot coexist with it, the two being mutually exclusive.' The two suggesters of the relation of cause and effect, in product and substance, are (1) concomitancy of affirmatives, that, while the product exists, the substance thereof exists, and (2) this concomitancy of negatives, that, when the substance no longer exists, the product no longer exists: and these two [conditions, on *your* theory] cannot be; because, since things [in your opinion,] are momentary in their duration, the two [*viz.*, the substance and the product], inasmuch as they are antecedent and consequent, belong to opposite times, [and cannot, therefore, coexist; for the product, according to you, does not come into existence until its substance has perished, which is contrary to the nature of the causal relation just defined].

b. But then, [the heretic may say, do not let the *coexistence* of substance and product be insisted upon, as indispensable to the causal relation between the two, but] 'let the nature of a cause belong to the substantial cause, as it belongs to the *instrumental* cause, in respect merely of its *antecedence*.' To this he replies:

Aph. 41.* If there were merely *antecedence*, then there would be no determination [of a substantial or material cause, as distinguished from an instrumental cause].

a. And it could not be determined that this was the *substance* [of this or that product], on the granting of nothing more than its *antecedence* [to the product]; because antecedence constitutes no distinction between it and the *instrumental* causes; for, [as we need scarcely remind you], that there *is* a distinction between instrumental and substantial causes, the whole world is agreed: such is the meaning.

b. Other heretics say: 'Since nothing [really] exists, except *Thought*, neither does *Bondage*; just as the things of a dream [have no real existence]. Therefore it has *no* cause; nor it is absolutely *false*.' He rejects the opinion of these [heretics]:

Aph. 42.* Not Thought alone exists; because there is the intuition of the external.

a. That is to say: the *reality* is not *Thought* alone; because external objects, also, are proved to exist, just as Thought is, by intuition.

b. But then [these heretics may rejoin], 'From the example of intuitive perception in *dreams* [see Butler's 'Analogy,' Part I., Ch. I.], we find this [your supposed evidence of objective reality] to exist, even in the *absence* of objects!' To this he replies:

Aph. 43.* Then, since, if the one does not exist, the other does not exist, there is a void, [*i.e.*, nothing exists at all].

a. That is to say: if external things do not exist, then a mere void offers itself. Why? Because, if the external does not exist, then *thought* does not exist; for it is *intuition* that proves the objective: and, if the intuition of the external did not establish the objective, then the intuition of *thought*, also, would not establish [the existence of] thought.

b. 'Then *let* the reality be a mere void; and, therefore, the searching for the cause of Bondage is unfitting, *just because a void is all*:' with such a proposal [as recorded in the next aphorism] does [some one who may claim the title of] the very crest-gem of the heretics rise up in opposition:

Aph. 44.* The reality is a void: what is perishes; because to perish is the habit of things.

a. The void alone [says this prince of heretics, or the fact that nothing exists at all] is the reality, [or the only truth]. Since everything that exists perishes, and that which is perishable is false, as is a dream, therefore, as of all things the beginnings and endings are merely nonentities, Bondage, &c., in the midst [of any beginning and ending], has merely a momentary existence,—is phenomenal, and not real. Therefore, *who* can be bound by *what*? This [question] is what we rest upon. The reason assigned for the perishableness of whatever exists is, 'because to perish is the habit of things;' because to perish is the *very nature* of things: but nothing continues, after quitting its own *nature*; [so that nothing could continue, if it *ceased* to perish]: such is the meaning.

b. He rejects [this heretical view]

Aph. 45.* This is a mere counter-assertion of unintelligent persons.

a. 'Of unintelligent persons,' *i.e.*, of blockheads, this is 'a mere counter-assertion,' *i.e.*, a mere *idle* counter-assertion that a thing must needs be perishable, *because it exists*; [and such an assertion is idle,] because things that are not made up of parts, since there is no cause of the destruction of such things, cannot perish.

b. [But] what need of many words? It is not the fact, that even *products* perish; [for] just as, by the cognition that 'the jar is old' [we mean that it has passed from the condition of new to that of old], so, too, by such a cognition as this, that 'the jar has passed away,' it is settled only that the jar, or the like, *is in the condition* of having passed away.

c. He states another objection [to the heretical view]:

Aph. 46.* Moreover, this [nihilistic theory is not a right one]; because it has the same fortune as both the views [which were confuted just before].

a. This view, moreover [§ 44], is not a good one; because it has the same fortune as, *i.e.*, is open to similar reasons for rejection as, the theory that external things are momentary [§ 26. *b*.], and as the theory that nothing exists besides Thought [§ 41. *b*.]. The reason for the rejection of the theory that things are momentary in their duration. *viz*. [as stated in § 35], the fact of *recognition*, &c., [which is, at least, as little consistent with Nihilism as it is with the momentary duration of things], and the reason for the rejection of the theory that nothing exists besides Thought. *viz*. [as stated in § 42], the intuition of the external, &c., apply equally here [in the case of Nihilism]: such is the import.

b. Moreover, as for the opinion which is accepted by these [heretics], *viz*., '*Let the mere void* [of absolute nonentity] be the soul's aim [and *summum bonum*], since herein consist at once the cessation of pain [which cannot continue, when there is absolutely *nothing*], and also the means thereof [since there can be no further means required for the removal of anything, if it be settled that the thing positively does not exist],' this too, can hardly be: so he declares [as follows]:

Aph. 47.* In neither way [whether as a means, or as an end,] is this [annihilation] the soul's aim.

a. 'Let the void [of mere nonentity] be the soul's aim, whether as consisting in the cessation of pain, or as presenting the means for the cessation of pain,' [says the heretic. And this cannot be; because the [whole] world agrees, that the aim of the soul consists in the joys, &c., that shall abide *in it*; that is to say, because [*they* hold, while] *you* do not hold, that there is a *permanent* soul, [(see § 33) in respect of which the liberation or beatification would be possible, or even predicable].

b. Now [certain] other things, also, entertained, as causes of [the soul's] bondage, by [imperfectly instructed] believers, remaining over and above those [proposed by unbelievers, and] already rejected, are to be set aside:

Aph. 48.* Not from any kind of motion [such as its entrance into a body, does the soul's bondage result].

a. 'Bondage' [required to complete the aphorism] is understood from the topic [of discussion].

b. The meaning is, that the soul's bondage, moreover, does not result from any sort of *motion*, in the shape, for instance, of its entrance into a body.

c. He states a reason for this:

Aph. 49.* Because this is impossible for what is inactive, [or in other words without motion]

a. That is to say: because this is impossible, *i.e.*, *motion* is impossible, in the case of the soul, which is inactive, [because] all-pervading, [and, therefore, incapable of changing its place].

b. But then [the objector may say], 'Since, in the books of Scripture and of law, we hear of its *going* and *coming* into this world and the other world, let soul be [not all-pervading, as you allege, but] merely limited [in its extent]: and to this effect, also, is the text, 'Of the size of the thumb is the soul, the inner spirit,' and the like: [but] this conjecture he repels:

Aph. 50.* [We cannot admit that the soul is other than all-pervading; because] by its being limited, since it would come under the same conditions as jars, &c., there would be a contradiction to our tenet [of its imperishableness].

a. That is to say: and, if the soul were admitted to be, like a jar, or the like, limited, *i.e.* circumscribed [in dimension], then, since it would resemble a jar, or the like, in being made up of parts, and [hence] in being perishable, &c., this would be contrary to our settled principle, [that the soul is imperishable].*

b. He now justifies the text [see § 49. *b.*] referring to the *motion* [of the soul, by showing that the motion is not really of the soul, but of an accessory]:

Aph. 51.* The text regarding the *motion* [of the soul], moreover, is [applicable, only] because of the junction of an *attendant*; as in the case of the Ether [or *Space*, which moves not, though we talk of the space enclosed in a jar, as moving with the jar].

a. Since there are such proofs of the soul's unlimitedness, as the declaration that 'It is eternal, omnipresent, permanent,' the text regarding its motion is to be explained as having reference to a movement pertaining [not to the soul, but] to an attendant; for there is the text, 'As the Ether [or space] included in a jar, when the jar is removed, [in this case] the *jar* may be removed, but not the space; and in like manner is the soul, which is like the sky, [incapable of being moved]'; and because we may conclude that the motion [erroneously supposed to belong to the soul (49. *b.*),] belongs to *Nature* [see *Vedanta Aphorisms*, Part I., § 4. *l.*], from such maxims as this, that 'Nature does the works the fruits of which are blissful or baneful; and it is wilful *Nature* that] in the three worlds, reaps these': such is the import.

b. It has already been denied [§ 16] that the bondage [of the soul] is occasioned by works] in the shape either of enjoined or of forbidden actions. Now he declares that the bondage, moreover] does not arise from the 'unseen' [merit or demerit] resulting therefrom:

Aph. 52.* Nor, moreover, [does the bondage of the soul result from the merit or demerit arising] from works; because these belong not thereto.

a. That is to say: the bondage of the soul does not arise directly from the 'unseen' [merit or demerit] occasioned by works. Why? Because this is no property thereof, *i.e.*, because this [merit or demerit (see § 16. *a*.)] is no property of the soul.

b. But then [some one may say], '*Let* it be that the bondage resulting from the 'unseen,' *i.e.*, the merit [or demerit] even of another, should attach to a different person;' whereupon he declares [as follows]:

Aph. 53.*4 If the case were otherwise [than as I say], then it [the bondage of the soul might extend unduly, even to the emancipated].

a. That is to say: if the case were otherwise, if bondage and its cause were under other conditions [than we have declared them to be], then there might be an undue extension; bondage would befall even the emancipated, [for the same reasons as those stated under § 16. *a*.].

b. What need of so much [prolixity]? He states a general objection why the bondage of soul cannot result from any one or other [of these causes], beginning with its essence [see § 6. *b*.], and ending with its [supposed] works [see § 16]; inasmuch as it is contrary to Scripture, [that any one of these should be the cause]:

Aph. 54.* And this [opinion, that the bondage of the soul arises from any of causes alleged by the heretics,] is contrary to such texts as the one that declares it [the soul] to be without qualities: and so much for that point.

a. And, if the bondage of the soul arose from any one or other of those [supposed causes already treated of,] among which its essential character [§ 6. *b*.] is the first, this would be contradictory to such texts as, 'Witness, intelligent, alone, and without the [three] qualities [is the soul]:' such is the meaning.

b. The expression 'and so much for that point' means, that the investigation of the cause of the bondage [of the soul] here closes.

c. The case, then, stands thus: since [all] other [theories] are overthrown by the declaratory aphorisms, 'There would be no fitness in the enjoining' [see § 7], &c., it is ascertained that the immediate cause of the bondage [of the soul] is just the conjunction of Nature and of the soul.

d. But then, in that case, [some one may say], this conjunction of Nature and of the soul [§ 54. *c*.], whether it be essential, or adventitiously caused by Time or something else [§ 5. *b*.], must occasion the bondage even of the *emancipated*. Having pondered this doubt, he disposes of it [as follows]:

Aph. 55.* Moreover, the conjunction thereof does not, through non-discrimination, take place [in the case of the emancipated]; nor is there a parity, [in this respect, between the emancipated and the unemancipated].

a. 'The conjunction thereof,' *i.e.*, the conjunction of Nature and of the soul; this conjunction, moreover, does not take place again 'through non-discrimination,' *i.e.*, through the want of a discrimination [between Nature and soul] in the emancipated, [who *do* discriminate, and who thus avoid the conjunction which others, failing to discriminate, incur, and thus fall into bondage]: such is the meaning. And thus the emancipated and the bound are *not* on a level, [under the circumstances stated at § 54. *c*.]: such is the import.

Aph. 56.* Bondage arises from the error [of not discriminating between Nature and soul].

a. Having thus declared the cause of that [bondage] which is to be got rid of, he declares the means of getting rid of it:]

Aph. 56.* The removal of it is to be effected by the necessary means, just like darkness.

a. The necessary means, established throughout the world, in such cases as 'shell-silver' [*i.e.*, a pearl-oyster-shell mistaken for silver], *viz.*, the *immediacy* of discrimination, by *this* alone is 'its removal,' *i.e.*, the removal of the non-discrimination [between Nature and soul], to be effected, and not by *works*, or the like: such is the meaning: just as darkness, the dark, is removed by light alone, [and by no other means].

b. 'But then [some one may say], if merely the non-discrimination of Nature and soul be, through the conjunction [of the two, consequent on the want of discrimination], the cause of bondage, and if merely the discrimination of the two be the cause of liberation, then there would be liberation, even while there remained the conceit of [one's possessing] a body, &c.; and this is contrary to Scripture, to the institutes of law, and to sound reasoning.' To this he replies:

Aph. 57.* Since the non-discrimination of other things [from soul] results from the non-discrimination of *Nature* [from soul], the cessation of this will take place, on the cessation of that [from which it results].

a. By reason of the non-discrimination of *Nature* from the soul, what non-discrimination of *other* things there *is*, such as the non-discrimination of the *understanding* [as something other than the soul], *this* necessarily ceases, on the cessation of the non-discrimination of Nature; because, when the non-discrimination of the understanding, for example, [as something other than soul,] does occur, it is *based* on the non-discrimination [from soul] of that cause to which there is none antecedent [*viz.*, Nature]; since the non-discrimination of an *effect* [and the 'understanding' is an effect or product of Nature,] is, itself, an effect, [and will, of course, cease, with the cessation of its cause].

b. The state of the case is this: as, when the soul has been discriminated from the *body*, it is impossible but that it should be discriminated from the colour and other [properties], the effects of the body, [which is the substantial cause of its own properties]; so, by parity of reasoning, from the departure of the cause, when soul, in its character of *unalterableness*, &c., has been discriminated from *Nature*, it is impossible that there should remain a conceit of [the soul's being any of] the *products* thereof [*i.e.*, of Nature], such as the 'understanding,' and the like, which have the character of being *modifications* [of primal

Nature, while the soul, on the other hand, is a thing unalterable].*

c. But then [some one may say], 'What proof is there that there is a conceit [entertained by people in general,] of a *Nature* [or primal principle] different from the conceit of an 'understanding,' &c., [which, you tell us, are products of this supposed first principle]? For all the various conceits [that the soul falls into], such as, 'I am ignorant,' and so on, can be accounted for on the ground simply of an 'understanding,' &c., [without postulating a primal Nature which is to assume the shape of an 'understanding,' &c.]:' well, if any one says this, I reply, 'no;' because, unless there were such a thing as Nature, we could not account for such conceits as the following, *viz.*, 'Having died, having died, again, when there is a creation, let me be a denizen of Paradise, and not of hell;' because no *products*, such as the 'understanding,' when they have perished, can be created anew, [any more than a gold-bracelet, melted down, can be reproduced, though another like it may be produced from the materials].

d. Moreover, it is inadmissible to say that men's conceit of [the identity of themselves with their] 'understanding,' &c., is [the *primary* cause of the soul's bondage, and is] not preceded by anything; because 'understanding' and the rest [as you will not deny] are *effects*. Now, while it is to be expected that there should be some predetermining agency to establish a conceit of [ownership in, or of one's identity with,] any *effects*, it is clear that it is a conceit of [ownership, &c.,] in respect of the *cause*, and nothing else, that must be the predetermining agency: for we see this in ordinary life; and our theories are bound to conform [deferentially] to experience. For [to explain,] we see, in ordinary life, that the conceit of [the ownership of] the grain, &c., produced by a field, results from the conceit of [the ownership of] the field; and, from the conceit of [the ownership of] gold, the conceit of [the ownership of] the bracelets, or other things, formed of that gold; and, by the removal of these [*i.e.*, the removal of the logically antecedent conceits, that the field, or the gold, is one's property], there is the removal of those, [*i.e.*, the removal of the conceits that the grain, &c., and that the bracelets, &c., the corresponding products or effects of the field and of the gold, are one's property: and so the soul will cease to confound itself with the 'understanding,' when it ceases to confound itself with Nature, of which the 'understanding' is held to be a product].

e. [And, if it be supposed that we thus lay ourselves open to the charge of a *regressus ad infinitum*, seeing that, whatever we may assign as the *first* cause, we may, on our own principles, be asked what was the 'predetermining agency' in regard to *it*; or if it be supposed thut we are chargeable with reasoning in a circle, when we hold that the soul's confounding itself with Nature is the cause of its *continuing* so to confound itself, and its continuing so to confound itself is, reciprocally, the cause why it confounds itself; we reply, that] there is no occasion to look for any other 'predetermining agency,' in the case of the conceit of [the identity of the soul with] Nature, or in the case of the self-continuance thereof, [*i.e.*, of that error of confounding one's self with Nature]; because [these two are alike] without antecedent, like seed and sprout, [of which it is needless to ask which is the first; the old puzzle, 'which was first, the acorn, or the oak?' being a frivolous question].

f. But then [some one may say], if we admit the soul's bondage [at one time], and its freedom [at another], and its discrimination [at one time], and its non-discriminatian [at another], then this is in contradiction to the assertion [in § 19], that it is 'ever essentially a pure and free intelligence;' and it is in contradiction to such texts as this, *viz.*, 'The absolute truth is this, that neither is there destruction [of the soul], nor production [of it]; nor is it bound, nor is it an effecter [of any work], nor is it desirous of liberation, nor is it, indeed, *liberated*; [seeing that that cannot desire or obtain liberation, which was never *bound*].' This [charge of inconsistency] he repels:

Aph. 58.* It is merely verbal, and not a reality [this so-called bondage of the soul]; since it [the bondage] *resides* in the *mind*, [and not in the soul].

a. That is to say: since bondage, &c., all reside only in the *mind* [and not in the soul], all this, as regards the soul, is merely verbal, *i.e.*, it is *vox et praeterea nihil*; because is is merely a *reflexion*, like the redness of [pellucid] crystal [when a China-rose is near it], but not a reality, with no false imputation, like the redness of the China-rose itself. Hence there is *no* contradiction to what had been said before, [as the objector (under § 57. *f.*) would insinuate]: such is the state of the case.

b. But then, if bondage, &c., as regards the soul, be merely verbal, let them be set aside by *hearing* [that they are merely verbal], or by argument [establishing that they are so]. Why, in the Scripture and the Law, is there enjoined, as the cause of liberation, a discriminative knowledge [of Soul, as distinguished from Non-soul], going the length of *immediate cognition*? To this he replies:

Aph. 59.* Moreover, it [the non-discrimination of Soul from Nature,] is not to be removed by argument; as that of a person perplexed about the points of the compass [is not to be removed] without immediate cognition.

a. By 'argument' we mean thinking. The word 'moreover' is intended to aggregate [or take in, along with 'argument'] 'testimony,' [or verbal authority, which, no more than 'argument,' or inference, can remove the evil, which can be removed by nothing short of direct intuitive *perception* of the real state of the case].

b. That is to say: the bondage, &c., of the soul though [granted to be] merely verbal, are not to be removed by merely hearing, by inferring, without immediate cognition, without directly perceiving; just as the contrariety in regard to the [proper] direction, though merely verbal [as resulting from misdirection], in the case of a person who is mistaken as to the points of the compass [and hence as to his own bearings], is not removed by testimony, or by inference, without immediate cognition, *i.e.*, without [his] directly perceiving [how the points of the compass really lie, to which immediate perception 'testimony,' or 'inference,' may conduce, but the necessity of which these *media*, or instruments of knowledge; cannot supersede].

c. Or it [Aph. 59] may be explained as follows, *viz*.: But then, [seeing that] it is declared, by the assertion [in Aph. 56], *viz*., that 'The removal of it is to be effected by the necessary means,' that knowledge, in the shape of discrimination [between soul and Nature], is the remover of *non*-discrimination [in regard to the matter in question], tell us, is that knowledge of a like nature with the hearing [of Testimony], &c.? Or is it something peculiar? A reply to this being looked for, he enounces the aphorism [§ 59]: 'Moreover, it is not to be removed by argument,' &c. That is to say: non-discrimination is not excluded, is not cut off, by argument, or by testimony, unless there be discrimination as an immediate perception; just as is the case with one who is bewildered in regard to [his] direction; because the only thing to remove an *immediate* error is an immediate individual perception [of the truth. For example, a man with the jaundice perceives *white* objects as if they were *yellow*. He may *infer* that the piece of chalk which he looks at is really white; or he may believe the *testimony* of a friend, that it *is* white; but still nothing will remove his erroneous *perception* of yellowness in the chalk, except a direct perception of whiteness.

d. Having thus, then, set forth the fact that Liberation results from the immediate discrimination [of Soul from Nature], the next thing to be set forth is the 'discrimination' [here referred to].

e. This being the topic, in the first place, since only if Soul and Nature exist, liberation can result from the discrimination of the one from the other, therefore that 'instrument of right knowledge' (*pramaIa*) which establishes the existence of these [two *imperceptible* realities] is [first] to be set forth:

Aph. 60.* The knowledge of things imperceptible is by means of Inference; as that of fire [when not directly perceptible,] is by means of smoke, &c.

a. That is to say: 'of things imperceptible,' *i.e.*, of things not cognizable by the senses, *e.g.*, Nature and the Soul, 'the knowledge,' *i.e.*, the fruit lodged in the soul, is brought about by means of that instrument of right knowledge [which may be called] 'Inference' (*anumana*), [but which (see Nyaya

Aphorisms, I., § 5) is, more correctly, 'the recognition of a Sign']; as [the knowledge that there is] fire [in such and such a locality, where we cannot directly perceive it,] is brought about by the 'recognition of a Sign,' occasioned by smoke, &c.

b. Moreover, it is to be understood that that which is [true, but yet is] not established by 'Inference,' is estalished by Revelation. But, since 'Inference' is the chief [among the instruments of knowledge], in this [the Samkhya] System, 'Inference' only is laid down [in the aphorism,] as the *chief* thing; but Revelation is not disregarded [in the Samkhya system; as will be seen from Aph. 88 of this Book].

c. He [next] exhibits the order of creation of those things among which Nature is the first, and the relation of cause and effect [among these, severally], preparatorily to the argument that will be [afterwards] stated:

Aph. 61.* Nature (*prak[iti*) is the state of equipoise of Goodness (*sattwa*) Passion (*rajas*), and Darkness (*tamas*): from Nature [proceeds] Mind (*mahat*); from Mind, Self-consciousness (*ahankara*); from Self-consciousness, the five Subtile Elements (*tan-matra*), and both sets [external and internal,] of Organs (*indriya*); and, from the Subtile Elements, the Gross Elements (*sthula-bhuta*). [Then there is] Soul (*purusha*). Such is the class of twenty-five.

a. 'The state of equipoise' of the [three] things called 'Goodness,' &c., is their being neither less nor more [one than another]; that is to say, the state of *not* being [developed into] an *effect* [in which one or other of them predominates]. And thus 'Nature' is the triad of 'Qualities' (*guIa*), distinct from the products [to which this triad gives rise]: such is the complete meaning.

b. These things, *viz*., 'Goodness,' &c., [though spoken of as the three Qualities], are not 'Qualities' (*guIa*) in the *Vaisheshika* sense of the word; because [the 'Qualities' of the *Vaisheshika* system have, themselves, *no* qualities (see KaGada's 16th Aph.); while] *these* have the qualities of Conjunction, Disjunction, Lightness, Force, Weight, &c. In this [Samkhya] system, and in Scripture, &c., the word 'Quality' (*guIa*) is employed [as the name of the three things in question], because they are subservient to Soul [and, therefore, hold a secondary rank in the scale of being], and because they form the *cords* [which the word *gula* also signifies], *viz.*, 'Mind,' &c., which consist of the three [so-called] 'Qualities,' and which *bind*, as a [cow, or other] brute-beast, the Soul.

c. Of this [Nature] the principle called 'the great one' (*mahat*), *viz.*, the principle of Understanding, (*buddhi*), is the product. 'Self-consciousness' is a conceit [of separate personality]. Of this there are two products, (1) the 'Subtile Elements' and (2) the two sets of 'Organs.' The 'Subtile Elements' are [those of] Sound, Touch, Colour, Taste, and Smell. The two sets of 'Organs,' through their division into the external and the internal, are of eleven kinds. The products of the 'Subtile Elements' are the five 'Gross Elements.' But 'Soul' is something distinct from either product or cause. Such is the class of twenty-five, the aggregate of things. That is to say, besides these there is nothing.

d. He [next], in [several] aphorisms, declares the order of the inferring [of the existence of these principles, the one from the other]:

Aph. 62.* [The knowledge of the existence] of the five 'Subtile Elements' is [by inference,] from the 'Gross Elements.'

a. 'The knowledge, by inference,' so much is supplied, [to complete the aphorism, from Aph. 60].

b. Earth, &c., the 'Gross Elements,' are proved to exist, by Perception; [and] thereby [*i.e.*, from that Perception; for Perception must precede Inference, as stated in Gotama's 5th Aphorism,] are the 'Subtile Elements' inferred. And so the application [of the process of inference to the case] is as follows:

(1) The Gross Elements, or those which have not reached the absolute limit [of simplification, or of the atomic], consist of things [Subtile Elements, or Atoms,] which have distinct qualities; [the earthy element having the distinctive quality of Odour; and so of the others]:

(2) Because they are gross;

(3) [And everything that is gross is formed of something less gross, or, in other words, more subtile,] as jars, webs, &c.; [the gross web being formed of the less gross threads; and so of the others].

Aph. 63.* [The knowledge of the existence] of Self-consciousness is [by inference,] from the external and internal [organs], and from these ['Subtile Elements,' mentioned in Aph. 62].

a. By inference from [the existence of] the external and internal organs, and from [that of] these 'Subtile Elements,' there is the knowledge of [the existence of such a principle as] Self-consciousness.

b. The application [of the process of inference to the case] is in the following [somewhat circular] manner:

(1) The Subtile Elements and the Organs are made up of things consisting of Self-consciousness:

(2) Because they are products of Self-consciousness:

(3) Whatever is not so [*i.e.*, whatever is *not* made out of Self-consciousness] is not thus [*i.e.*, is not a *product* of Self-consciousness]; as the Soul, [which, not being made up thereof, is not a product of it].

c. But then, if it be thus [*i.e.*, if it be, as the Sankhyas declare, that all objects, such as jars, are made up of Self-consciousness, while Self-consciousness depends on Understanding,' or 'Intellect,' or 'Mind,' the *first* product of Nature' (see Aph. 61)], then [some may object, that], since it would be the case that the Self-consciousness of the potter is the material of the jar, the jar made by him would disappear, on the beatification of the potter, whose internal organ [or 'Understanding'] then surceases. And this [the objector may go on to say,] is not the case; because *another* man [*after* the beatification of the potter,] recognizes that 'This is that same jar [which, you may remember, was fabricated by our deceased acquaintance].'

d. [In reply to this we say,] it is *not* thus; because, on one's beatification, there is an end of only those modifications of his internal organ [or 'Intellect'] which could be causes [as the *jar* no longer can be,] of the emancipated soul's *experiencing* [either good or ill], but not an end of the modifications of intellect in general, nor [an end] of intellect altogether: [so that we might spare ourselves the trouble of further argument, so far as concerns the objection grounded on the assumption that the intellect of the potter *surceases*, on his beatification: but we may go further, and admit, for the sake of argument, the surcease of the 'intellect' of the beatified potter, without conceding any necessity for the surcease of his pottery. This alternative theory of the case may be stated as follows]:

e. Or [as Berkeley suggests, in his Principles of Human knowledge, Ch. vi.], let the Self-consciousness of the *Deity* be the cause why jars and the like [continue to exist], and not the Self-consciousness of the potter, &c., [who may lose their Self-consciousness, whereas the Deity, the sum of all life, *HiraIyagarbha* (see *Vedanta-shara*, § 62), never loses *his* Self-consciousness, while aught living continues].

Aph. 64.* [The knowledge of the existence] of Intellect is [by inference,] from that [Self-consciousness, § 63].

a. That is to say: by inference from [the existence of] 'that,' *viz.*, Self-consciousness, which is a product, there comes the knowledge of 'Intellect' (*buddhi*), the *great* 'inner organ' (*anta-karaIa*), [hence] called 'the great one' (*mahat*), [the existence of which is recognized] under the character of the *cause* of this [product, *viz.*, Self-consciousness].

b. And so the application [again rather circular, of the process of inference to the case,] is as follows:

(1) The thing called Self-consciousness is made out of the things that consist of the moods of judgment [or mind];

(2) Because it is a thing which is a product of judgment [proceeding in the Cartesian order of *cogito, ergo sum*; and]

(3) Whatever is not so [*i.e.*, whatever is *not* made out of judgment, or mental assurance], is not thus [*i.e.*, is not a product of mental assurance]; as the Soul, [which is not made out of this or of anything antecedent], &c.

c. Here the following reasoning is to be understood: Every one, having first determined anything under a concept [*i.e.*, under such a form of thought as is expressed by a general term; for example, that this which presents itself is a jar, or a human body, or a possible action of one kind or other], after that makes the judgment, 'This is I,' or 'This ought to be done by me,' and so forth: so much is quite settled; [and there is no dispute that the fact is as here stated]. Now, having, in the present instance, to look for some *cause* of the thing called 'Self-consciousness' [which manifests itself in the various judgments just referred to], since the relation of cause and effect subsists between the two functions [the occasional conception, and the subsequent occasional judgment, which is a function of Self-consciousness], it is assumed, for simplicity, merely that the relation of cause and effect exists between the two substrata to which the [two sets of] functions belong; [and this is sufficient,] because it follows, as a matter of course, that the occurrence of a *function* of the effect must result from the occurrence of a *function* of the cause; [nothing, according to the Samkhya, being in any product, except so far, and in such wise, as it preexisted in the cause of that product].

Aph. 65.* [The knowledge of the existence] of Nature is [by inference,] from that ['Intellect,' § 64].

a. By inference from [the existence of] 'that,' *viz*., the principle [of Intellect, termed], 'the Great one,' which is a *product*, there comes the knowledge of [the existence of] Nature, as [its] cause.

b. The application [of the process of inference to the case] is as follows:

(1) Intellect, the affections whereof are Pleasure, Pain, and Dulness, is produced from something which has these affections, [those of] Pleasure, Pain, and Dulness:

(2) Because, whilst it is a *product* [and must, therefore, have arisen from something consisting of that which itself now consists of], it consists of Pleasure, Pain, and Dulness; [and]

(3) [Every *product* that has the affections of, or that occasions, Pleasure, Pain, or Dulness, takes its rise in something which consists of these]; as lovely women, &c.

c. For an agreeable woman gives pleasure to her husband, and, therefore, [is known to be mainly made up of, or] partakes of the quality of 'Goodness;' the indiscreet one gives pain to him, and, therefore, partakes of the quality of 'Foulness;' and she who is separated [and perhaps forgotten,] occasions indifference, and so partakes of the quality of Darkness.'

d. And the appropriate refutation [of any objection], in this case, is [the principle], that it is fitting that the qualities of the effect should be [in every case,] in conformity with the qualities of the cause.

e. Now he states how, in a different way, we have [the evidence of] inference for [the existence of] Soul, which is void of the relation of cause and effect that has been mentioned, [in the four preceding aphorisms, as existing between Nature and its various products]:

Aph. 66.* [The existence] of Soul [is inferred] from the fact that the combination [of the principles of Nature into their various effects] is for the sake of another [than unintelligent Nature, or any of its similarly unintelligent products].

a. 'Combination,' *i.e.*, conjunction, which is the cause [of all products; these resulting from the conjunction of their constituent parts]. Since whatever has this quality, as Nature, Mind, and so on [unlike soul, which is *not* made up of parts], is for the sake of some other; for this reason it is understood that soul exists: such is the remainder, [required to complete the aphorism].

b. But the application [of the argument, in this particular case, is as follows]:

(1) The thing in question, *viz.*, Nature the 'Great one,' with the rest [of the aggregate of the unintelligent], has, as its fruit [or end], the [mundane] experiences and the [eventual] Liberation of some other than itself:

(2) Because it is a combination [or *compages*];

(3) [And every combination,] as a couch, or a seat, or the like, [is for another's use, not for its own; and its several component parts render no mutual service].

c. Now, in order to establish that it is the cause of all [products], he establishes the *eternity* of Nature (*prak[iti*):

Aph. 67.* Since the root has no root, the root [of all] is rootless.

a. Since 'the root' (*mula*), *i.e.*, the cause of the twenty-three principles, [which, with soul and the root itself, make up the twenty-five realities recognized in the Samkhya,] 'has no root,' *i.e.*, has no cause, the 'root,' *viz.*, Nature (*pradhana*), is 'rootless,' *i.e.*, void of root. That is to say, there is no other cause of Nature; because there would be a *regressus in infinitum*, [if we were to suppose another cause, which, by parity of reasoning, would require another cause; and so on without end].

b. He states the argument [just mentioned] in regard to this, [as follows]:

Aph. 68.* Even if there be a succession, there is a halt at some *one* point; and so it is merely a name [that we give to the point in question, when we speak of the *root* of things, under the the name of 'Nature'].

a. Since there would be the fault of *regressus in infinitum*, if there were a succession of causes,—another cause of Nature, and another [cause] of that one, again,—there must be, at last, a halt, or conclusion, at some one point, somewhere or other, at some one, uncaused, eternal thing. Therefore, that at which we stop is the *Primal Agency* (*pra-k[iti*); for this [word *prak[iti*, usually and conveniently rendered by the term *Nature*,] is nothing more than a sign to denote the cause which is the *root*: such is the meaning.

b. But then [some Vedant*i* may object according to this view of matters], the position that there are just twenty-five realities is not made out; for, in addition to the 'Indiscrete' [or primal Nature], which [according to you,] is the cause of Mind, *another* unintelligent principle, named 'Ignorance' [see *Vedanta-sara*, § 21], presents itself. Having pondered this doubt, he declares [as follows]:

Aph. 69.* Alike, in respect of Nature, [and of both Soul and Nature, is the argument for the uncreated existence].

a. In the discussion of the Primal Agent [Nature], the cause which is the root [of all products], the same side is taken by us both, the asserter [of the Samkhya doctrine] and the opponent [Vedant*i*]. This may be thus stated: As there is mention, in Scripture, of the *production* of Nature, so, too, is there of that of *Ignorance*, in such texts as this, *viz*.: 'This Ignorance, which has five divisions, was produced from the great Spirit.'* Hence it must needs be that a figurative production is intended to be asserted, in respect of *one* of these [and not the *literal* production of both; else we should have no root at all]; and, of the two, it is with *Nature* only that a figurative production, in the shape of a manifestation through conjunction with Soul, &c., is congruous. A production [such as that metaphorical one here spoken of,] the characteristic of which is conjunction *is* mentioned; for there is mention of [such] a figurative origination of Soul and Nature, in a passage of the *Kaurma* [*Purala*], beginning, 'Of action [or the Primal Agency], and knowledge [or Soul],' and so on. And, as there is no mention, in Scripture, of the origin of *Ignorance*, as figurative, *it* is *not* from eternity. And Ignorance, which consists of false knowledge, has been declared, in an aphorism of the *Yoga*, to be [not a separate entity, but] 'an affection of the mind.' Hence there is no increase to the [list of the twenty-five] Realities, [in the shape of a twenty-sixth principle, to be styled ignorance].

b. Or [according to another, and more probable, interpretation of the aphorism,] the meaning is this, that the argument is the same in support of both, *i.e.*, of both Soul and Nature: such is the meaning.

c. But then, there being [as has been shown,] a mode of arriving, by inference, at [a knowledge of the saving truth in regard to] Nature, Soul, &c., whence is it that reflexion, in the shape of discrimination [between Soul and Nature], does not take place in the case of *all* [men]? In regard to this point, he states [as follows]:

Aph. 70.* There is no rule [or necessity, that *all* should arrive at the truth]; because those who are privileged [to engage in the inquiry] are of three descriptions.

a. For those privileged [to enguge in the inquiry] are of three descriptions, through their distinction into those who, in reflecting, are dull, mediocre, and best. Of these, by the dull the [Samkhya] arguments are frustrated [and altogether set aside], by means of the sophisms that have been uttered by the *Bauddhas*, &c. By the mediocre they [are brought into doubt, or, in other words,] are made to appear as if there were equally strong arguments on the other side, by means of arguments which really prove the reverse [of what these people employ them to prove], or by arguments which are not true: [see the section on Fallacies in the *Tarka-sangraha*]. But it is only the *best* of those privileged, that reflect in the manner that has been set forth [in our exposition of the process of reflexion which leads to the discriminating of Soul from Nature]: such is the import. But there is no rule that *all* must needs reflect in the manner so set forth: such is the literal meaning.

b. He now, through two aphorisms, defines 'the Great one' and 'Self-consciousness'; [the reader being presumed to remember that Nature consists of the three 'Qualities' in equipoise, and to be familiar with the other principles, such as the 'Subtile elements' (see § 61)]:

Aph. 71.* The first product [of the Primal Agent, Nature], which is called 'the Great one,' is Mind.

a. 'Mind' (*manas*). 'Mind' [is so called], because its function is 'thinking' (*manana*). By 'thinking' is here meant 'judging' (*nishchaya*). That of which this is the function is 'intellect' (*buddhi*); and *that* is the first product, that called 'the Great one' (*mahat*): such is the meaning.

Aph. 72.* 'Self-consciousness' is that which is subsequent [to Mind.]

a. 'Self-consciousness,' the function of which is a conceit [that '*I* exist,' '*I* do this, that, and the other thing'], is that which is subsequent: that is to say, 'Self-consciousness' is the next after 'the Great one' [§ 71].

b. Since 'Self-consciousness' is that whose function is a conceit [which brings out the *Ego*, in every case of cognition, the matter of which cognition would, else, have lain dormant in the bosom of Nature, the formless Objective], it therefore follows that the others [among the phenomena of mundane existence,] are effects of this [Self-consciousness]; and so he declares [as follows]:

Aph. 73.* To the others it belongs to be products thereof, [*i.e.*, of Self-consciousness].

a. 'To be products thereof,' *i.e.*, to be products of Self-consciousness: that is to say, the fact of being products thereof belongs to the others, the eleven 'Organs' (*indriya*), the five 'Subtile elements,' and, mediately, to the [gross] Elements, also, the products of the Subtile elements.

b. But then, if it be thus [some one may say], you relinquish your dogma, that Nature is the cause of the whole world. Therefore he declares [as follows]:

Aph..74.* Moreover, mediately, through that [*i.e.*, the 'Great one' (§ 71)] the first [cause, *viz.*, Nature,] is the cause [of all products]; as is the case with the Atoms, [the causes, though not the immediate causes, of jars, &c.].

a. 'Moreover, mediately,' *i.e.*, moreover, not in the character of the immediate cause, 'the first,' *i.e.*, Nature, is the Cause of 'Self-consciousness' and the rest, [mediately,] through 'the Great one' and the rest; as, in the theory of the *Vaisheshikas*, the Atoms are the cause of a jar, or the like, only [mediately,] through combinatians of two atoms, and so on: such is the meaning.

b. But then, since, also, both Nature and Soul are eternal, which of them is [really] the cause of the creation's commencing? In regard to this, he declares [as follows]:

Aph. 75.* While both [Soul and Nature] are antecedent [to all products], since the one [*viz.*, Soul,] is devoid [of this character of being a cause], it is applicable [only] to the other of the two, [*viz.*, Nature].

a. That is to say: 'while both,' *viz.*, Soul and Nature, are preexistent to every product, still, 'since the one,' *viz.*, Soul, from the fact of its not being modified [into anything else, as clay is modified into a jar], must be 'devoid,' or lack the nature of a cause, 'it is applicable,' *i.e.*, the nature of a cause must belong, to the *other* of the two.

b. But then [some one may say], let *Atoms* alone be causes; since there is no dispute [that *these* are causal]. In reply to this, he says:

Aph. 76.* What is limited cannot be the substance of all [things].

a. That which is limited cannot be the substance of all [things]; as yarn cannot be the [material] cause of a jar. Therefore it would [on the theory suggested,] be necessary to mention separate causes of [all] things severally; and it is simpler to assume a single cause. Therefore Nature alone is the cause. Such is the meaning.

b. He alleges Scripture in support of this:

Aph. 77.* And [the proposition that Nature is the cause of all is proved] from the text of Scripture, that the origin [of the world] is therefrom, [*i.e.*, from Nature].

a. An argument, in the first instance, has been set forth [in § 76; for, till argument fails him, no one falls back upon authority]. Scripture, moreover, declares that Nature is the cause of the world, in such terms as, 'From Nature the world arises,' &c.

b. But then [some one may say], a jar which antecedently did not exist is seen to come into existence. Let, then, *antecedent non-existence* be the cause [of each product]; since this is an invariable antecedent, [and, hence, a cause; 'the invariable antecedent being denominated a cause,' if Dr. Brown, in his 6th lecture, is to be trusted]. To this he replies:

Aph. 78.* A thing is not made out of nothing.

a. That is to say: it is not possible that out of nothing, *i.e.*, out of a nonentity, a thing should be made, *i.e.*, an entity should arise. If an entity were to arise out of a nonentity, then, since the character of a cause is visible in its product, the *world*, also, would be unreal: such is the meaning.

b. Let the world, too, *be* unreal: what harm is that to us? [If any ask this,] he, therefore, declares [as follows]:

Aph. 79.* It [the world] is not unreal; because there is no fact contradictory [to its reality], and because it is not the [false] result of depraved causes, [leading to a belief in what ought not to be believed].

a. When there is the notion, in regard to a shell [of a pearl-oyster, which sometimes glitters like silver], that it is silver, its being silver is contradicted by the [subsequent and more correct] cognition, that this is *not* silver. But, in the case in question [that of the world regarded as a reality], no one ever has the cognition, 'This world is *not* in the shape of an entity,' by which [cognition, if any one ever really had such,] it being an entity might be opposed.

b. And it is held that that is false which is the result of a *depraved* cause; *e.g.*, some one's cognition of a [white] conch-shell as *yellow*, through such a fault as the jaundice, [which depraves his eye-sight]. But, in the case in question, [that of the world regarded as a reality], there is not such [temporary or occasional] depravation [of the senses]; because all, at all times, cognize the world as a reality. Therefore the world is *not* an unreality.

c. But then [some one may suggest], *let* a nonentity be the [substantial] cause of the world; still the world will not [necessarily, therefore,] be unreal. In regard to this, he declares [as follows]:

Aph. 80.* If it [the substantial cause,] be an entity, then this would be the case, [that the product would be an entity], from its union [or identity] therewith; [but] if [the cause be] a nonentity, then how could it possibly be the case [that the product would be real], since *it* is a nonentity, [like the cause with which it is united, in the relation of identity]?

a. If an *entity* were the substantial cause [of the world], then, since [it is a maxim that] the qualities of the cause present themselves in the product, 'this would be the case,' *i.e.*, it would be the case that the product was real, 'because of union therewith,' *i.e.*, because of the union [of the product] with the reality [which is its subatratum]. [But,] since, [by parity of reasoning], if a *nonentity* [were the substantial cause], the world would be a nonentity, then, by reason of its being a nonentity, *i.e.*, by reason of the world's being [on that supposition,] necessarily a non-entity, [like its supposed cause], how could this be the case, [that it would be *real*]?

b. But then [a follower of the *MimaEsa* may say], since [it would appear that] nonentity can take no shape but that of nonentity, let *works* alone be the cause of the world. What need have we of the hypothesis of 'Nature'? To this he replies:

Aph. 81.* No; for *works* are not adapted to be the *substantial* cause [of any product].

a. Granting that 'the uneeen' [merit or demerit arising from actions] may be an *instrumental* cause, [in bringing about the mundane condition of the agent], yet we never see merit or demerit in the character of the *substantial* cause [of any product]: and our theories ought to show deference to our experience. 'Nature' is to be accepted; because Liberation arises [see § 56,3 and § 83,] from discerning the distinction between Nature and the Soul.

b. But then [some one may say], since Liberation can be attained by undertaking the things directed by the Veda, what occasion is there for [our troubling ourselves about] *Nature*? To this he replies:

Aph. 82.* The accomplishment thereof [*i.e.*, of Liberation] is not moreover, through Scriptural rites: the chief end of man does not consist in this [which is gained through such means]; because, since this consists of what is accomplished through *acts*, [and is therefore, a *product*, and not *eternal*], there is [still left impending over the ritualist,] the liability to repetition of births.

a. 'Scriptural means,' such as sacrifices, [are so called], because they are heard from [the mouth of the instructor in] Scripture. Not thereby, moreover, is 'the accomplishment thereof,' *i.e.*, the accomplishment of Liberation; 'because one is liable to repetition of births, by reason of the fact that it [the supposed Liberation,] was accomplished by *means*,' *i.e.*, because the [thus far] liberated [soul] is still liable to repetition of births, inasmuch as this [its supposed Liberation,] is not *eternal*,

[just] because it is [the result of] *acts*. For *this* reason, the chief end of man does not consist in this, [which is gained through ritual observances].

b. He shows what *does* constitute the chief end of man:

Aph. 83.* There is Scripture for it, that he who has attained to discrimination, in regard to these [*i.e.* Nature and Soul], has no repetition of births.

a. 'In regard to these,' *i.e.*, in regard to Nature and Soul, of him who has attained to discrimination, there is a text declaring, that, in consequence of his knowledge of the distinction, there will be no repetition of births; the text, *viz.*, '*He* does not return again,' &c.

b. He states an objection to the opposite view:

Aph. 84.* From pain [occasioned, *e.g.*, to victims in sacrifice] must come pain [to the sacrificer, and not *liberation* from pain]; as there is not relief from chilliness, by affusion of water.

a. If Liberation were to be effected by *acts*, [such as sacrifices], then, since the acts involve a variety of pains, Liberation itself [on the principle that every effect includes the qualities of its cause,] would include a variety of pains; and it would be a grief, from the fact that it must eventually end: for, to one who is distressed by chilliness the affusion or water does not bring liberation from his chilliness, but, rather, [additional] chilliness.

b. But then [some one may say], the fact that the act is productive of pain is not the *motive* [to the performance of sacrifice]; but the [real] reason is this, that the act is productive of *things desirable*. And, in accordance with this, there is the text, 'By means of acts [of sacrifice] they may partake of immortality,' &c. To this he replies:

Aph. 85.* [Liberation cannot arise from acts]; because whether the end be something desirable, or undesirable, [and we admit that the *motive* of the sacrifice is not the giving pain to the victim], this makes no difference in regard to its being the result of *acts*, [and, therefore, not eternal, but transitory].

a. Grant that pain is not what is [intended] to be accomplished by works done without desire, [on the part of the virtuous sacrificer], still, though there *is* a difference [as you contend,] between [an act done to secure] something enjoyable and an act done without reference to enjoyment, this makes no difference with repect to the fact of the Liberation's being produced by *acts*, [which, I repeat, *permanent* Liberation cannot be]: there must still again be pain; for it [the Liberation supposed to have been attained through works,] must be perishable, because it is a *production*. The text which declares that works done without desire are instruments of Liberation has reference to *knowledge*, [which, I grant, may be gained by such means]; and Liberation comes through knowledge; so that these [works] are instruments of Liberation *mediately*: [but you will recollect that the present inquiry regards the *immediate* cause].

b. [But then, some one may say], supposing that Liberation may take place [as you Sankhyas contend,] through the knowledge of the distinction between Nature and Soul, still, since, from the perishableness [of the Liberation effected by *this* means, as well as any other means], mundane life may return, were both on an equality, [*we*, whose Liberation you Sankhyas look upon as transitory, and you Sankhyas, whose Liberation we, again, look upon as being, by parity of reasoning, in much the same predicament]. To this he replies:

Aph. 86.* Of him who is essentially liberated, his bonds having absolutely perished, it [*i.e.*, the fruit of his saving knowledge,] is absolute: there is no parity [between his case and that of him who relies on works, and who may thereby secure a temporary sojourn in Paradise, only to return again to earth].

a. Of him 'who is essentially liberated,' who, in his very essence, is free, there is the destruction of bondage. The bond [see § 56,1] is Non-discrimination [between Nature and Soul]. By the removal thereof there is the destruction, the annihilation, of Non-discrimination: and how is it possible that there should again be a return of the mundane state, when the destruction of Non-discrimination is *absolute*? Thus there is no [such] similarity, [between the two cases, as is imagined, by the objector, under § 85. *b*.].

b. It has been asserted [in § 61,] that there is a class of twenty-five [things which are realities], and, since these cannot be ascertained [or made out to be *true*], except by *proof*, therefore he displays this; [*i.e.*, he shows what he *means* by proof]:

Aph. 87.* The determination of something not [previously] lodged in both [the Soul and the Intellect], nor in one or other of them, is 'right notion' (*prama*). What is, in the highest degree, productive thereof [*i.e.*, of any given 'right notion'], is that; [*i.e.*, is what we mean by proof, or evidence, (*pramaIa*)].

a. 'Not lodged,' *i.e.*, not deposited in 'one rightly cognizing' (*pramat[i*); in short, not previously known. The 'determination,' *i.e.*, the ascertainment [or right comprehension] of such a thing, or reality, is 'right notion'; and, whether this be an affection 'of *both*,' *i.e.*, of Intellect, and also of Soul [as some hold that it is], or of only one or other of the two, [as others hold,] *either* way, 'what is, in the highest degree, productive' of this 'right notion' is [what we term proof, or] evidence, (*pramaIa*): such is the definition of evidence in general; [the definition of its several species falling to beconsidered hereafter]: such is the meaning.

b. It is with a view to the exclusion of Memory, Error, and Doubt, in their order, that we employ [when speaking of the result of evidence,] the expressions 'not previously known' [which excludes things remembered], and 'reality' [which excludes mistakes and fancies], and 'discrimination,' [which excludes doubt].

c. In regard to this [topic of knowledge and the sources of knowledge], if 'right notion,' is spoken of as located in the *Soul* [see § 87. *a*.], then the [proof, or] evidence is an affection of the *Intellect*. If [on the other hand, the 'right notion' is spoken of as] located in the Intellect, in the shape of an affection [of that the affections of which are mirrored by the Soul], then it [the proof, or evidence, or whatever we may choose to call that from which 'right notion' results,] is just the conjunction of an organ [with its appropriate object; such conjunction giving rise to sense-perception], &c. But, if *both* the Soul's cognition and the affections of the Intellect are spoken of as [cases of] 'right notion,' then *both* of these aforesaid [the affection of the Intellect, in the first case, and the conjunction of an organ with its appropriate object, &c., in the other case,] are [to receive the name of] proof (*pramaIa*). You are to understand, that, when the organ of vision, &c., are spoken of as 'evidence,' it is only as being *mediately* [the sources of right knowledge].

d. How many [kinds of] proofs [then,] are there? To this he replies:

Aph. 88.* Proof is of three kinds:1 there is no establishment of more; because, if these be established, then all [that is true] can be established [by one or other of these three proofs].

a. 'Proof is of three kinds;' that is to say, 'perception' (*pratyaksha*), 'the recognition of signs' (*anumana*), and 'testimony' (*shabda*), are the [three kinds of] proofs.

b. But then [some one may incline to say], let 'comparison' [which is reckoned, in the Nyaya, a specifically distinct source of knowledge], and the others [such as 'Conjecture,' &c., which are reckoned, in like manner, in the M*i*maEsa], also be instruments of right knowledge, [as well as these three], in [the matter of] the discriminating of Nature and Soul: he therefore says, 'because, if those [three] be established,' &c. And, since, if there be the three kinds of proof established, 'everything [that is really true] can be established [by means of them], there is no establishment of more;' no addition to the proofs can be fairly made out; because of the cumbrousness [that sins against the philosophical maxim, that we are not to assume more than is necessary to account for the case]: such is the meaning.

c. For the same reason, Manu, also, has laid down only a triad of proofs, where he says [see the Institutes, Ch. xii., v. 105]: 'By that man who seeks a distinct knowledge of his duty, [these] three [sources of right knowledge] must be well understood, *viz.*, Perception, Inference, and Scriptural authority in its various shapes [of legal institute, &c.].' And 'Comparison,' and 'Tradition' (*aitihya*), and the like, are included under Inference and Testimony; and 'Non-perception' (*anupalabdhi*) and the like are included under Perception; [for the non-perception of an absent jar on a particular spot of ground is nothing else than the perception of that spot of ground *without* a jar on it].

d. He [next] states the definitions of the varieties [of proof, having already (§ 87) given the general definition]:

Aph. 89.* Perception (*pratyakska*) is that discernment which, being in conjunction [with the thing perceived], portrays the form thereof.

a. 'Being in conjunction,' [literally,] 'existing in conjunction;' 'portrays the form thereof,' *i.e.*, assumes the form of the thing with which it is in conjunction [as water assumes the form of the vessel into which it is poured]; what 'discernment,' or affection of the Intellect, [does *this*], that [affection of the Intellect (see Yoga Aphorisms, I., § 5 and § 8. *b*.)] is the evidence [called] Perception: such is the meaning.

b. But then, [some one may say,] this [definition of Perception (§ 89)] does not extend [as we conceive it ought, and presume it is intended, to do,] to the perception, by adepts in the *Yoga*, of things past, future, or concealed [by stone walls, or such intervening things as interrupt ordinary perception]; because there is, here, 'no form of the thing, in *conjunction*' [with the mind of him who perceives it, while absent]: having pondered this doubt, he corrects it by [stating, as follows,] the fact, that this [supernatural sort of perception] is not what he intends to define:

Aph. 90.* It is not a fault [in the definition, that it does not apply to the perceptions of adepts in the *Yoga*]; because that of the adepts in the *Yoga* is not an *external* perception.

a. That is to say: it is only *sense*-perception that is to be here defined; and the adepts of the Yoga do not perceive through the *external* [organs of sense]. Therefore there is no fault [in our definition]; *i.e.*, there is no *failure* to include the perceptions of these; [because there is no *intention* to include them].

b. [But, although this reply is as much as the objector has any right to expect,] he states the real justification [of the definition in question]:

Aph. 91.* Or, there is no fault [in the definition], because of the conjunction, with *causal* things, of that [mystical mind] which has attained exaltation.

a. Or, be it so that the perception of the *Yogi*, also, shall be the thing to be defined; still there is no fault [in our definition, § 89]; it does not fail to extend [to this, also]; since the mind of the *Yogi*, in the exaltation gained from the habitude produced by concentration, *does* come into conjunction with things [as existent] in their causes, [whether or not with the things as developed into products perceptible by the external senses].

b. Here the word rendered 'causal' (*lina*) denotes the things, *not* in conjunction [with the senses], alluded to by the objector [in § 89. *b*.]; for *we*, who assert that effects *exist* [from eternity, in their causes, before taking the shape of effects, and, likewise, in these same causes, when again resolved into their causes], hold that even what is past, &c., still essentially exists, and that, hence, its conjunction [with the mind of the mystic, or the clairvoyant,] is possible.

c. But then, [some one may say,] still this [definition] does not extend to the *Lord's* perceptions; because, since these are from everlasting, they cannot *result* from [emergent] conjunction. To this he replies:

Aph. 92.* [This objection to the definition of Perception has no force]; because it is not proved that there *is* a Lord (*ishwara*).

a. That there is no fault [in the definition of Perception], because there is no proof that there *is* a Lord, is supplied [from § 90].

b. And this demurring to there being any 'Lord' is merely in accordance with the arrogant dictum of [certain] partisans [who hold an opinion not recognized by the majority]. Therefore, it is to be understood, the expression employed is, 'because it is *not proved* that there is a Lord,' but not the expression, 'because there *is no* Lord,'

c. But, on the implication that there *is* a 'Lord,' what we mean to speak of [in our definition of Perception, (§ 89),] is merely the being of the [same] kind with what is produced by conjunction [of a sense-organ with its object; and the perceptions of the 'Lord' may be of the same *kind* with such perceptions, though they were not to come from the same *source*].

d. Having pondered the doubt, '*How* should the Lord not be proved [to exist] by the Scripture and the Law, [which declare his existence]?' he states a dilemma which excludes [this]:

Aph. 93.* [And, further,] it is not proved that he [the 'Lord,'] exists; because [whoever exists must be either free or bound; and], of free and bound, he can be neither the one nor the other.

a. The 'Lord' whom you imagine, tell us, is he free from troubles, &c.? Or is he in bondage through these? Since he is not, cannot be, either the one or the other, it is not proved that there is a 'Lord:' such is the meaning.

b. He explains this very point:

Aph. 94.* [Because,] either way, he would be inefficient.

a. Since, if he were free, he would have no desires, &c., which [as compulsory motives,] would instigate him to create; and, if he were bound, he would be under delusion; he must be [on either alternative,] unequal to the creation, &c. [of this world].

b. But then, [it may be asked,] if such be the case, what becomes of the Scripture-texts which declare the 'Lord?' To this he replies:

Aph. 95.* [The Scriptural texts which make mention of the 'Lord' are either glorifications of the liberated Soul, or homages to the recognized [deities of the Hindu pantheon].

a. That is to say: accordingly as the case may be, *some* text [among those in which the term 'Lord' occurs,] is intended, in the shape of a glorification [of Soul], as the 'Lord,' [as Soul is held to be], merely in virtue of junction [with Nature], to incite [to still deeper contemplation], to exhibit, as what is to be known, the liberated Soul, *i.e.*, absolute Soul in general; and some other text, declaratory, for example, of creatorship, &c., preceded by resolution [to create, is intended] to extol [and to purify the mind of the contemplator, by enabling him to take a part in extolling] the eternity, &c., of the familiarly known Brahma, VishIu, *sh*iva, or other *non*-eternal 'Lord;' since these, though possessed of the conceit [of individuality], &c., [and, in so far, liable to perish], have immortality, &c., in a secondary sense; [seeing that the Soul, in *every* combination, is immortal, though the combination itself is not so].

b. But then, [some one may say], even if it were thus [as alleged under § 95], what is heard in Scripture, [*viz.*], the fact that it [*viz.*, Soul] is the *governor* of Nature, &c., would not be the case; for, in the world, we speak of government in reference only to modifications [preceded and determined] by resolutions [that so and so shall take place], &c. To this he replies:

Aph. 96.* The governorship [thereof, *i.e.*, of Soul over Nature] is from [its] proximity thereto, [not from its resolving to act thereon]; as is the case with the gem, [the lodestone, in regard to iron].

a. If it were alleged that [its, Soul's] creativeness, or [its] governorship, was through a *resolve* [to create, or to govern], then this objection [brought forward under § 95. *b*.] would apply. But [it is not so; for,] by *us* [Sankhyas,] it is held that the Soul's governorship, in the shape of creatorship, or the like, is merely from [its] *proximity* [to Nature]; 'as is the case with the [lodestone] gem.'

b. As the gem, the lodestone, is attracted by iron merely by proximity, without resolving [either to act or to be acted on], &c., so, by the mere conjunction of the primal Soul, Nature is changed into the principle [called] the 'Great one,' [or Mind, (see § 61. *c*.)]. And in this alone consists [what we speak of as] its acting as *creator* towards that which is superadded to it: such is the meaning

c. And thus it is declared, [in some one of the PuraIas]: 'As the iron acts, whilst the gem [the lodestone,] stands void of volition, just so this world is created by a deity who is mere Existence. Thus it is, that there are, in the Soul, both agency [seemingly,] and non-agency, [really]. It is *not* an agent, inasmuch as it is void of volition; [and it *is*] an agent, merely through approximation [to Nature].'

d. In respect of worldly products, also, animal souls overrule, merely through their approximation [to Nature]: so he declares [as follows]:

Aph. 97.* In the case of individual products, also, [the apparent agency] of animal souls [is solely through proximity].

a. 'The agency is solely through proximity:' so much is supplied [from § 96].

b. The meaning is this, that, in the case, also, of particular productions,—the creation, &c., of things individual [as contradistinguished from that of all things in the lump, (see *Vedanta-sara*, § 67)],—animal souls, *i.e.*, souls in which the intellects [of individuals] reflect themselves [see § 99. *a*.], overrule, merely through proximity, but not through any effort; seeing that these [animal souls] are none other than the motionless Thought.

c. But then, [some one may say], if there were no eternal and omniscient 'Lord,' through the doubt of a blind tradition, [in the absence of an intelligently effective guardianship], the *Vedas* would cease to be an authority; [a possibility which, of course, cannot be entertained for an instant]. To this he replies:

Aph. 98.* The declaration of the texts or sense [of the Veda, by Brahma, for example], since *he* knows the truth, [*is* authorative evidence].

a. To complete [the aphorism, we must say], 'since *HiraIyagarbha* [*i.e.*, *Brahma*,] and others [*viz.*, *VishIu* and *shiva*], are knowers of what is certain, *i.e.*, of what is true, the declaration of the texts or sense of the Vedas, where *these* are the speakers, *is* evidence [altogether indisputable].'

b. But then, if Soul, by its simple proximity [to Nature (§ 96)], is an overruler in a *secondary* sense [only of the term,—as the magnet may be said, is a secondary sense, to draw the iron, while the conviction is entertained, that, actually and literally, the iron draws the magnet],—who is the *primary* [or actual,] overruler? In reference to this, he says:

Aph. 99.* The internal organ, through its being enlightened thereby [*i.e.*, by Soul], is the overruler; as is the iron, [in respect of the magnet].

a. The internal organ, *i.e.*, the understanding, is the overruler, through its fancying itself to be Soul, [as it does fancy,] by reason of its being enlightened by the Soul, through its happening to reflect itself in [and contemplate itself in,] Soul; 'just as the iron,' that is to say, as the attracting iron, though inactive, draws [the magnet], in consequence of [its] mere proximity, [and so acquires magnetism by magnetic induction].

b. He [now, having discussed the evidence that consists in direct perception,] states the definition of inference (*anumana*):

Aph. 100.* The knowledge of the connected [*e.g.*, fire], through perception of the connexion [*e.g.*, of fire with smoke], is inference.

a. That is to say: inference [or conviction of a general truth,] is [a kind of] evidence consisting in a [mental] modification, [which is none other than] the knowledge of the connected, *i.e.*, of the constant accompanier, through the knowledge of the constant accompaniment: by 'connexion' (*pratibandha*) here being meant 'constant attendedness' (*vyapti*); and through the perception thereof [it being that the mind has possession of any general principle].

b. But a conclusion (*anumiti*) is knowledge of the soul; [whilst an Inference, so far forth as it is an instrument in the establishment of knowledge deducible from it, is an affection of the internal organ, or understanding (see § 87. *c*.)]

c. He [next] defines testimony (*shabda*):

Aph. 101.* Testimony [such as is entitled to the name of evidence, is a declaration by one worthy [to be believed].

a. Here 'fitness' means 'suitableness;' and so the evidence which is called 'Testimony' is the knowledge arising from a suitable declaration: such is the meaning. And [while this belongs to the understanding, or internal organ (see § 100. *b*.)] the result is that [knowledge] in the Soul, [which is called,] 'knowledge by hearing' (*shabda-bodha*).

b. He [next] volunteers to tell us what is the use of his setting forth [the various divisions of] evidence:

Aph. 102.* Since the establishment of [the existence of] both [soul and non-soul] is by means of evidence, the decaration thereof [*i.e.*, of the kinds of evidence, has been here made].

a. It is only by means of evidence that both Soul and non-soul are established as being distinct, [the one from the other]: therefore has this, *viz.*, evidence, been here declared: such is the meaning.

b. Among these [several kinds of proof], he [now] describes that one by which, especially, *viz.*, by a proof which is one kind of inference, Nature and Soul are here to be established discriminatively:

Aph. 103.* The establishment of both [Nature and Soul] is by analogy.

a. [Analogy (*samanyato d[ishma*) is that kind of evidence which is employed in the case] where, by the force [as an argument,] which the residence of any property in the subject derives from a knowledge of its being constantly accompanied [by something which it may therefore betoken], when we have had recourse to [as the means of determining this constant accompaniment,] what is, for instance, generically of a perceptible kind, [where, under such circumstances, we repeat,] anything of a *different* kind, *i.e.*, *not* cognizable by the senses, is established; as when, for example, having apprehended a constant accompaniment, [*e.g.*, that an act implies an instrument], by taking into consideration such instruments as axes, &c., which are of earthy and other kinds, a quite heterogeneous, imperceptible, instrument of *knowledge*, *viz.*, [the instrument named] Sense, is established [or inferred to exist]; such is what we mean by Analogy; and it is by *this* [species of inference], that both, [*viz.*,] Nature and Soul are proved [to exist]: such is the meaning.

b. Of these [*viz.*, Nature and Soul,] the argument from analogy for [the existence of] Nature is as follows: the Great Principle [*viz.*, Understanding (see § 61. *c.*)] is formed out of the things [called] Pleasure, Pain, and Delusion, [to the aggregate of which three in equipoise (see § 61) the name of Nature is given]; because, whilst it is [undeniably,] a production, it has the characters of Pleasure, Pain, and Delusion; just as a bracelet, or the like, formed of gold, or the like; [has the characteristic properties of the gold, or the like, and is thereby known to have been formed out of gold, or the like].

c. But, [as regards tht argument from analogy, in proof of the existence] of Soul, [it is, as stated before, under § 66, to the following effect]: Nature is for the sake of *another*; because it is something that acts as a combination; as a house, for instance, [which is a combination of various parts combined for the benefit of the tenant]. In this instance, having gathered, in regard to houses, &c., the fact established on sense-perception, that they exist for the sake of [organized] bodies, for example, something of a different kind therefrom, [*i.e.*, from Nature, *viz.*], Soul, is inferred [by analogy,] as something other than Nature, &c., [which, as being a compound thing, is not designed for itself]: such is the meaning.

d. But then [some one may say], since Nature is eternal, and exertion is habitual to her, [and the result of her action is the bondage of the Soul], there should constantly be experience [whether of pleasure or of pain], and, hence, no such thing as thorough emancipation. To this he replies:

Aph. 104.* Experience [whether of pain or pleasure, ends with the discernment of] Thought, [or Soul, as contradistinguished from Nature].

a. By 'Thought' [we mean] soul. Experience [whether of pain or pleasure,] ceases, on the discerning thereof. As 'antecedent non-existence,' though devoid of a beginning, [see *Tarka-sangraha*, § 92], surceases [when the thing antecedently non-existent begins to be], so, eternal Nature [eternal, as regards the absence of any beginning,] continues [no further than] till the discernment of the difference [between Nature and Soul]; so that experience [whether of pain or pleasure,] does *not* at all times occur: such is the state of the case.

b. [But some one say], if Nature be agent, and Soul experiencer, then it must follow [which seems unreasonable,] that another is the experiencer of [the results of] the acts done by one different. To this he replies:

Aph. 105.* The experience of the fruit *may* belong even to another than the agent; as in the case of food, &c.

a. As it belongs to the cook to prepare the food, &c., and to one who was not the agent, *viz.*, the master, to enjoy the fruit [thereof, *i.e.*, the fruit of the cook's actions], so is the case here, also.

b. Having stated an exoteric principle [which may serve, in practice, to silence, by the *argumentum ad hominem*, him on whose principles it may be valid], he [next] declares his own doctrine, [in regard to the doubt started under § 104. *b*.]:

Aph. 106.* Or, [to give a better account of the matter than that given in § 105], since it is from non-discrimination that it is derived, the notion that the *agent* [soul being mistaken for an agent,] has the fruit [of the act is a wrong notion].

a. The soul is neither an agent nor a patient; but, from the fact that the Great Principle [the actual agent (see § 97. *b*.)] is reflected in it, there arises the *conceit* of its being an agent. 'Or, since it is from non-discrimination;' that is to say, because it is from the failure to discriminate between Nature and Soul, that this takes place, *i.e.*, that conceit takes place, that it is the *agent* that experiences the fruit; [whereas the actual agent is Nature, which, being unintelligent, can experience neither pain nor pleasure].

b. The opposite of this [wrong view, referred to in § 106,] he states [as follows]:

Aph. 107.* And, when the truth is told, there is [seen to be] neither [agency, in Soul, nor experience].

a. 'When the truth is told' [and discerned], *i.e.*, when, by means of evidence, Nature and Soul are *perceived* [in their entire distinctness, one from the other], 'there is neither,' *i.e.*, neither the condition [as regards soul,] of an agent nor that of a patient.

b. Having discussed [the topic of] evidence, he [now] states the distribution of the subject-matter of evidence:

Aph. 108.* [A thing may be] an object [perceptible], and also [at another time,] not an object, through there being, in consequence of great distance, &c., a want of [conjunction of the sense with the thing], or [on the other hand,] an appliance of the sense [to the thing].

a. An object [is a perceived object], through the proximity, or conjunction, of the sense [with the object]. [A thing may be] not an object [perceived], through the want of the sense, *i.e.*, through the want of conjunction [between the sense and what would otherwise be its object]. And [this] want of conjunction [may result] from the junction's being prevented by great distance, &c.

b. [To explain the '&c.,' and to exemplify the causes that may prevent the conjunction, required in order to perception, between the thing and the sense, we may remark, that] it is in consequence of great distance, that a bird [flying very high up] in the sky is not perceived; [then again,] in consequence of extreme proximity, the collyrium located in the eye [is not perceived by the eye itself]; a thing placed in [the inside of, or on the opposite side of,] a wall [is not perceived], in consequence of the obstruction; from distraction of mind, the unhappy, or other [agitated person], does not perceive the thing that is at his side [or under his very nose]; through its subtilty, an atom [is not perceived]; nor is a very small sound, when overpowered by the sound of a drum; and so on.

c. How [or, for which of the possible reasons just enumerated,] comes the imperceptibleness of *Nature*? In regard to this, he declares:

Aph. 109.* Her imperceptibleness arises from [her] subtility.

a. 'Her,' *i.e.*, *Nature's*, imperceptibleness is from subtilty. By subtilty is meant the fact of being difficult to investigate; not [as a Naiyayika might, perhaps, here prefer understanding the term,] the consisting of atoms; for Nature is [not atomic, in the opinion of the Sankhyas, but] all-pervasive.

b. How, then, [it may be asked,] is [the existence of] Nature determined? To this he replies:

Aph. 110.* [Nature exists;] because her existence is gathered from the beholding of productions.

a. As the knowledge of [there being such things as] atoms comes from the beholding of jars, &c., [which are agglomerations], so the knowledge of Nature comes from the beholding of products which have the three Qualities; [(see § 62. *a.*) and the existence of which implies a cause, to which the name of Nature is given, in which these constituents exist from eternity].

b. Some [the Vedant*is*,] say that the world has *Brahma* as its cause; others [the Naiyayikas], that it has atoms as its cause; but our seniors [the transmitters of the Samkhya doctrine], that it has *Nature* as its cause. So he sets forth a doubt [which might naturally found itself] thereon:

Aph. 111.* If [you throw out the doubt that] it [*viz.*, the existence of Nature,] is not established, because of the contradiction of asserters [of other views, then you will find an answer in the next aphorism].

a. 'Because of the the contradiction of asserters [of the Vedanta or Nyaya], *it* is not established,' *i.e.*, Nature [as asserted by the Sankhyas,] is not established.

b. But then, [to set forth the objection of these counter-asserters], if a product existed antecedently to its production [as that product], *then* an eternal Nature [such as you Sankhyas contend for,] would be proved to exist as the [necessary] substratum thereof; since you will declare that a cause is inferred only as the [invariable] accompanier of an effect; but it is denied, by us asserters [of the Vedanta, &c.], that the effect *does* exist [antecedently to its production; well,] *if* [this doubt be thrown out]: such is the meaning [of the aphorism].

c. He states [his] doctrine [on this point]:

Aph. 112.* Still, since each [doctrine] is established in the opinion of each, a [mere unsupported] denial is not [decisive].

a. If one side were disproved merely by the dissent of the opponent, then [look you,] there is dissent against the other side, too: so how could *it* be established? If the one side is established by there being inevitably attendant the recognition of the constant accompanier, on the recognition of that which is constantly accompanied [by it], it is the same with *my* [side], also: therefore [my] inference from effect [to cause] is not to be denied [in this peremptory fashion].

b. Well, then, [the opponent may say], let [the inference of] cause from effect be granted; how is it that this [cause] is *Nature*, and nothing else, [such as Atoms, for instance]? To this he replies:

Aph. 113.* Because [if we were to infer any other cause than Nature,] we should have a contradiction to the threefold [aspect which things really exhibit].

a. Quality is threefold [see § 61. *a*.], *viz*., Goodness, Passion, and Darkness: there would be a contradiction to *these*: such is the meaning.

b. The drift here is as follows: If the character of cause [of all things around us] belonged to Atoms, or the like, then there would be a contradiction to the fact of being an aggregate of pleasure, pain, and delusion, which is recognizable in the world; [because nothing, we hold, can exist in the effect, which did not exist in the cause, and pleasure, pain, &c., are no properties of Atoms].

c. He now repels the doubt as to whether the production of an effect is that of what existed [antecedently], or of what did not exist:

Aph. 114.* The production of what is no entity, as a man's horn, does not take place.

a. Of that which, like the horn of a man, is not an entity, even the production is impossible: such is the meaning. And so the import is, that that effect alone which [antecedently] exists is [at any time] produced.

b. He states an argument why an effect must be some [previously existent] entity:

Aph. 115.* Because of the rule, that there must be some material [of which the product may consist].

a. And only when both are extant is there, from the presence of the cause, the presence of the effect. Otherwise, everywhere and always, every [effect] might be produced; [the presence of the cause being, on the supposition, superfluous]. This he insists upon [as follows]:

Aph. 116.* Because everything is not possible everywhere and always [which might be the case] if materials could be dispensed with].

a. That is to say: because, in the world, we see that everything is *not* possible, *i.e.*, that everything is *not* produced; 'everywhere,' *i.e.*, in every place; 'always,' *i.e.*, at all times.

b. For the following reason, also, he declares, there is no production of what existed not [antecedently]:

Aph. 117.* Because it is that which is competent [to the making of anything] that makes what is possible, [as a product of it].

a. Because the being the material [of any future product] is nothing else than the fact of [being it, *potentially*, *i.e.*, of] having the competency to be the product; and [this] competency is nothing else than the product's condition as that of what has not yet come to pass: therefore, since 'that which is competent,' *viz.*, the cause, makes the product which is 'possible' [to be made out of it], it is not of any *nonentity* that the production takes place, [but of an entity, whose *esse*, antecedently, was *possibility*]: such is the meaning.

b. He states another argument:

Aph. 118.* And because it [the product,] is [nothing else than] the cause, [in the shape of the product].

a. It is declared, in Scripture, that, previously to production, moreover, there is no difference between the cause and its effect; and, since it is thereby settled that a product is an entity, production is not of what [previously] existed not: such is the meaning.

b. He ponders a doubt:.

Aph. 119.* If [it be alleged that] there is no possibility of that's *becoming* which already *is* [then the answer will be found in the next aphorism].

a. That is to say: but then, if it be thus [that every effect exists antecedently to its production], since the effect [*every* effect,] must be eternal [without beginning], there is no possibility of [or room for] the adjunction of *becoming*, the adjunction of *arising*, in the case of a product which is [already, by hypothesis,] in the shape of an entity; because the employment of [the term] '*arising*' [or the fact of being produced] has reference solely to what did *not* exist [previously]; if this be urged: such is the meaning.

b. He declares the doctrine [in regard to this point]:

Aph. 120.* No; [do not argue that what *is* cannot become; for] the employment and the non-employment [of the term 'production'] are occasioned by the *manifestation* [and the non-manifestation of what is spoken of as produced, or not].

a. 'No;' the view stated [in § 119] is not the right one: such is the meaning.

b. As the whiteness of white cloth [which has become] dirty is brought manifestly out by means of washing, &c., so, by the operation of the potter, is the pot brought into manifestness; [whereas], on the blow of a mallet, it becomes hidden, [and no longer appears as a *pot*].

c. And manifestation [is no fiction of ours; for it] is seen; for example, that of oil, from sesamum-seeds, by pressure; of milk, from the cow, by milking; of the statue, which resided in the midst of the stone, by the operation of the sculptor; of husked rice, from rice in the husk, by threshing; &c.

d. Therefore, the employment and the non-employment of the [term] 'the *production* of an effect' are dependent on *manifestation*, dependent on the manifestation of the *effect*: that is to say, the employment of [the term] 'production' is in consequence of the manifestation [of what is spoken of as produced]; and the non-employment of [the term] 'production' is in consequence of there being no manifestation [of that which is, therefore, not spoken of as produced]; but [the employment of the term 'production' is] not in consequence of that's becoming an entity which was not an entity.

e. But if [the employment of the term] 'production' is occasioned by [the fact of] *manifestation*, by what is occasioned [the employment of the term] *destruction*? To this he replies:

Aph. 121.* Destruction [of anything] is the resolution [of the thing spoken of as destroyed,] into the cause [from which it was produced].

a. The resolution, by the blow of a mallet, of a jar into its cause [*i.e.*, into the particles of clay which constituted the jar], to *this* are due both [the employment of] the term 'destruction,' and the kind of action [or behaviour] belonging to anything [which is termed its destruction].

b. [But some one may say], if there were [only] a resolution [of a product into that from which it arose], a resurrection [or À±»133µ½µÃ¯±] of it might be seen; and this is *not* seen: well [we reply], it is not seen by blockheads; but it *is* seen by those who can discriminate. For example, when thread is destroyed, it is changed into the shape of earth [as when burned to ashes]; and the earth is changed into the shape of a cotton-tree; and this [successively] changes into the shape of flower, fruit, and thread [spun again from the fruit of the cotton-plant]. So is it with all entities.

c. Pray [some one may ask], is [this] *manifestation* [that you speak of under § 120] something real, or something not real? If it be something real [and which, therefore, never anywhere ceases to be], then [all] effects [during this constant manifestation] ought constantly to be *perceived*; and, if it be *not* real, then there would be the absence of [all] products, [in the absence of all manifestation. Manifestation, therefore, must be something *real*; and] there must be [in order to give rise to it,] another manifestation of it, and of this another; [seeing that a *manifestation* can be the result of nothing else than a manifestation, on the principle that an effect consists of neither more nor less than its cause]; and thus we have a *regressus in infinitum*. To this he replies:

Aph. 122.* Because they seek each other reciprocally, as is the case with seed and plant, [manifestation may generate manifestation, from eternity to eternity].

a. Be it so, that there are thousands of manifestations; still there is no fault; for there *is* no starting-point; as is the case with seed and plant, [which people may suppose to have served, from eternity, as sources, one to another, reciprocally].

b. He states another argument:

Aph. 123.* Or, [at all events, our theory of 'manifestation' is as] blameless as [your theory of] 'production.'

a. Pray [let us ask], is *production* produced, or is it not? If it is produced, then of this [production of production] there must be production; so that there is a *regressus in infinitum*, [such as you allege against *our* theory, under § 121. *c*.] If it be *not* produced, then, pray, is this because it is *unreal*, or because it is eternal? If because it is unreal, then production never is at all; so that it would never be perceived, [as you allege that it is]. Again, if [production is not something produced,] because it is *eternal*, then there would be at, all times, the production of [all possible] effects, [which you will scarcely pretend is the case]. Again, if you say, since 'production' itself *consists* of production, what need of supposing an ulterior production [of production]? then, in like manner, [*I* ask,] since 'manifestation' itself *consists* of manifestation, what need of supposing an ulterior manifestation [of manifestation]? The view which you hold on this point is *ours*, also; [and thus every objection stated or hinted under § 121. *c*., is capable of being retorted].

b. He [now] states the community of properties [that exists] among the products of Nature, mutually:

Aph. 124.* [A product of Nature is] caused, uneternal, not all-pervading, mutable, multitudinous, dependent, mergent.

a. 'Caused,' *i.e.* having a cause. 'Uneternal,' *i.e.*, destructible. 'Not all-pervading,' *i.e.*, not present everywhere. 'Mutable,' *i.e.*, distinguished by the acts of leaving [one form], and assuming [another form], &c. It [the soul,] leaves the body it has assumed, [and, probably, takes another]; and bodies, &c., move [and are mutable, as is notorious]. 'Multitudinous,' *i.e.*, in consequence of the distinction of souls; [every man, *e.g.*, having a separate body]. 'Dependent,' [*i.e.*,] on its cause. 'Mergent,' that is to say, it [*i.e.*, every product, in due time,] is resolved into that from which it originated.

b. [But, some one may say], if realities be the twenty-five [which the Sankhyas enumerate (see § 61), and no more], pray, are such common operations as knowing, enjoying, &c., absolutely *nothing*; you accordingly giving up what you *see*, [in order to save an hypothesis with which what you see is irreconcilable]? To this he replies:

Aph. 125.* There is the establishment of these [twenty-four 'Qualities' of the *Nyaya*, which you fancy that we do not recognize, because we do not explicitly enumerate them], either by reason that these ordinary qualities [as contradistinguished from the *three* Qualities of the Samkhya], &c., are, in reality, nothing different; or [to put it in another point of view,] because they are hinted by [the term] Nature, [in which, like our own three Qualities, they are implied].

a. Either from their being nothing different from the twenty-four principles, 'in reality,' truly, quite evidently,—since the character of these [twenty-four] fits the ordinary qualities, &c., [which you fancy are neglected in our enumeration of things,]—'there is the establishment of these,' *i.e.*, there is their establishment [as realities,] through their being implied just in those [twenty-four principles which are explicitly specified in the Samkhya].

b. The word 'or' shows that there is another alternative [reply, in the aphorism, to the objection in question]. 'Or because they are hinted by [the term] Nature;' that is to say, the qualities, &c. [such as Knowledge], are established [as realities], just because they are hinted by [the term] Nature, by reason that [these] qualities are, mediately, products of Nature; for there is no difference between product and cause. But the omission to mention them [explicitly] is not by reason of their not being at all.

c. He [next] mentions the points in which Nature and [her] products agree:

Aph. 126.* Of both [Nature and her products] the fact that they consist of the three Qualities [§ 61. *a*.], and that they are irrational, &c., [is the common property].

a. Consisting of the three qualities, and being irrational, [such in the meaning of the compound term with which the aphorism commences]. By the expression '&c.' is meant [their] being intended for *another*, [see § 66]. 'Of both,' *i.e.*, of the cause [*viz.*, Nature], and of the effects [*viz.*, all natural products]. Such is the meaning:

b. He [next] states the mutual differences of character among the three Qualities which [see § 61] are the [constituent] parts of Nature:

Aph. 127.* The Qualities [§ 62] differ in character mutually by pleasantness, unpleasantness, lassitude, &c., [in which forms, severally, the Qualities present themselves].

a. 'Pleasantness,' *i.e.*, Pleasure. By the expression '&c.' is meant Goodness (*sattwa*), which is light [*i.e.*, not heavy,] and illuminating. 'Unpleasantness,' *i.e.*, Pain. By the expression '&c.' [in reference to this,] is meant Passion (*rajas*), which is urgent and restless. 'Lassitude,' *i.e.*, stupefaction. By the expression '&c.' is meant Darkness (*tamas*), which is heavy and enveloping. It is by these habits that the Qualities, *viz.*, Goodness, Passion, and Darkness, differ: such is the remainder, [required to complete the aphorism].

b. At the time of telling their differences, he tells in what respects they agree:

Aph. 128.* Through Lightness and other habits the Qualities mutually agree and differ.

a. The meaning is as follows: the enunciation [in the shape of the term *laghu*, 'light,' is not one intended to call attention to the concrete, *viz.*, what things are light, but] is one where the abstract [the nature of light things, *viz.*, 'lightness' (*laghutwa*)] is the prominent thing. 'Through Lightness and other habits,' *i.e.*, through the characters of Lightness, Restlessness, and Heaviness, the Qualities differ. Their *agreement* is through what is hinted by the expression 'and other.' And this consists in their mutually predominating [one over another, from time to time], producing one another, consorting together, and being reciprocally present, [one in another], for the sake of Soul.

b. By [the expressions, in § 124,] 'caused,' &c., it is declared that the 'Great one' [or Mind], &c., are *products*. He states the proof of this:

Aph. 129.* Since they are other than both [Soul and Nature, the only two uncaused entities], Mind and the rest are products; as is the case with a jar, or the like.

a. That is to say: like a jar, or the like, Mind and the rest are products; because they are something other than the two which [alone] are eternal, *viz*., Nature and Soul.

b. He states another reason:

Aph. 130.* Because of [their] measure, [which is a limited one].

a. That is to say: [Mind and the rest are products]; because they are limited in measure; [whereas the only two that are uncaused, *viz*., Nature and Soul, are unlimited].

b. He states another argument:

Aph. 131.* Because they conform [to Nature].

a. [Mind and the rest are products]; because they well [follow and] correspond with Nature; *i.e.*, because the Qualities of Nature [§ 61] are seen in all things: [and it is a maxim, that what is in the effect was derived from the cause and implies the cause].

b. He states the same thing, [in the next aphorism]:

Aph. 132.* And, finally, because it is through the power [of the cause alone, that the product can do aught].

a. It is by the power of its cause, that a product energizes, [as a chain restrains an elephant, only by the force of the iron which it is made of]; so that Mind and the rest, being [except through the strength of Nature,] powerless, produce *their* products in subservience to Nature. Otherwise, since it is their habit to energize, they would at all times produce their products, [which it will not be alleged that they do].

b. And the word *iti*, in this place, is intended to notify the completion of the set of [positive] reasons [why Mind and the others should be regarded as products].

c. He [next] states [in support of the same assertion,] the argument from negatives, [*i.e.*, the argument drawn from the consideration as to what becomes of Mind and the others, when they are *not* products]:

Aph. 133.* On the quitting thereof [quitting the condition of product], there is Nature, or Soul, [into one or other of which the product must needs have resolved itself].

a. Product and non-product; such is the pair or alternatives. 'On the quitting thereof;' *i.e.*, when Mind and the rest quit the condition of product, Mind and the rest [of necessity] enter into Nature, or Soul; [these two alone being non-products].

b. [But perhaps some one may say, that] Mind and the rest may exist quite independently of the pair of alternatives [just mentioned]. In regard to this, he declares [as follows]:

Aph. 134.* If they were other than these two they would be void; [seeing that there is nothing self-existent, besides soul and Nature].

a. If Mind and the rest were 'other than these two,' *i.e.*, than product or non-product [§ 133], they would be in the shape of what is 'void,' *i.e.*, in the shape of nonentity.

b. Well now, [some one may say,] why should it be under the character of a *product*, that Mind and the rest are a sign of [there being such a principle as] Nature? They may be [more properly said to be] a sign, merely in virtue of their not *occurring apart* from it. To this he replies:

Aph. 135.* The cause is inferred from the effect, [in the case of Nature and her products]; because it accompanies it.

a. That [other relation, other than that of material and product, which you would make out to exist between Nature and Mind,] exists, indeed, where the nature [or essence] of the cause is not seen in the effect; as [is the case with] the inference, from the rising of the moon, that the sea is swollen [into full tide; rising, with maternal affection, towards her son who was produced from her bosom on the occasion of the celebrated Churning of the Ocean. Though the swelling of the tide does not occur apart from the rising of the moon, yet here the cause, moon-rise, is not seen in the effect, tide; and, consequently, though we infer the effect from the cause, the cause could not have been inferred from the effect]. But, in the present case, since we see, in Mind and the rest, the characters of Nature, the cause *is* inferred from the effect. 'Because it accompanies it,' *i.e.*, because, in Mind and the rest, we see the properties of Nature, [*i.e.*, Nature herself actually present; as we see the clay which is the cause of a jar, actually present in the jar].

b. [But it may still be objected,] if it be thus, then let that principle itself, the 'Great one' [or Mind], be the cause of the world: what need of *Nature*? To this he replies:

Aph. 136.* The indiscrete, [Nature, must be inferred] from its [discrete and resolvable] effect. [Mind], in which are the three Qualities, [which constitute Nature].

a. 'It is resolved;' such is the import of [the term] *linga*, [here rendered] 'effect.' From that [resolvable effect], *viz.*, the 'Great principle' [or Mind], in which are the three Qualities, Nature must be inferred. And that the 'Great principle,' in the shape of ascertainment [or distinct intellection], is discrete [or limited] and perishable, is established by direct observation. Therefore [*i.e.*, since Mind, being perishable, must be resolvable into something else,] we infer that into which it is resolvable, [in other words, its 'cause,' here analogously termed *lingin*, since 'effect' has been termed *linga*].

b. But then, [some one may say], still something quite different may be the cause [of all things]: what need of [this] *Nature* [of yours]? In regard to this, he remarks [as follows]:

Aph. 137.* There is no denying that it [Nature,] *is*; because of its effects, [which will be in vain attributed to any other source].

a. Is the cause of this [world] a product, or not a product? If it were a product, then, the same being [with equal propriety to be assumed to be] the case with *its* cause, there would be a *regressus in infinitum*. If effects be from any *root* [to which there is nothing antecedent], then *this* is that [to which we give the name of *Nature*]. 'Because of its effects,' that is to say, because of the effects of Nature. There is no denying 'that it is,' *i.e.*, that *Nature* is.

b. Be it so, [let us grant,] that Nature *is*; yet [the opponent may contend,] *Soul* positively cannot be; for [if the existence of causes is to be inferred from their products, Soul cannot be thus demonstrated to exist, seeing that] it has *no* products. In regard to this, he remarks [as follows]:

Aph. 138.* ['The relation of cause and effect is] not [alleged as] the means of establishing [the existence of Soul]; because, as is the case with [the disputed term] 'merit,' there is no dispute about there being such a kind of thing; [though *what* kind of thing *is* matter of dispute].

a. There is no dispute about 'there being such a kind of thing,' *i.e.*, as to there being Soul, simply; [since everybody who does not talk stark nonsense must admit a Soul, or *self*, of *some* kind]; for the dispute is [not as to its *being*, but] as to its peculiarity [of being], as [whether it be] multitudinous, or sole, all-pervading, or *not* all-pervading, and so forth; just as, in every [philosophical system, or] theory, there is no dispute as to [there being something to which may be applied the term] 'merit' (*dharma*); for the difference of opinion has regard to the particular kind of [thing,—such as sacrifices, according to the M*i*maEsa creed, or good works, according to the Nyaya,—which shall be held to involve] 'merit.'

b. 'Not the means of establishing' that [*viz.*, the existence of soul]; *i.e.*, the relation of cause and effect is not the means of establishing it. This intends, 'I will mention *another* means of establishing it.'

c. [But some one may say,] Souls are nothing else than the body, and its organs, &c.: what need of imagining anything else? To this he replies:

Aph. 139.* Soul is something else than the body, &c.

a. [The meaning of the aphorism is] plain.

b. He propounds an argument in support of this:

Aph. 140.* Because that which is combined [and is, therefore, discerptible,] is for the sake of some other, [*not* discerptible].

a. That which is discerptible is intended for something else that is indiscerptible. If it were intended for something else that is discerptible, there would be a *regressus ad infinitum.*

b. And combinedness [involving (see § 67) discerptibleness,] consists in the Qualities' making some product by their state of mutual commixture; or [to express it otherwise,] combinedness is the state of the soft and the hard, [which distinguishes matter from spirit]. And this exists occultly in Nature, as well as the rest; because, otherwise, discerptibleness would not prove discoverable in the products thereof, *viz.*, the 'Great one,' &c.

c. He elucidates this same point:

Aph. 141.* [And Soul is something else than the body, &c.]; because there is [in Soul,] the reverse of the three Qualities, &c.

a. Because there is, in Soul, 'the reverse of the three Qualities,' &c., *i.e.*, because they are not seen [in it]. By the expression '&c.' is meant, because the *other* characters of Nature, also, are not seen [in soul].

b. He states another argument:

Aph. 142.* And [Soul is not material;] because of [its] superintedence [over Nature].

a. For a superintendent is an intelligent being; and Nature is unintelligent: such is the meaning.

b. He states another argument:

Aph. 143.* [And Soul is not material;] because of [its] being the experiencer.

a. It is Nature that is experienced; the experiencer is Soul. Although Soul, from its being unchangeably the same, is not [really] an experiencer, still the assertion [in the aphorism,] is made, because of the fact that the reflexion of the Intellect befalls it, [and thus makes it *seem* as if it experienced (see § 58. *a.*)].

b. Efforts are engaged in for the sake of Liberation. Pray, is this [for the benefit] of the Soul, or of Nature; [since Nature, in the shape of Mind, is, it seems, the experiencer]? To this be replies:

Aph. 144.* [It is for Soul, and not for Nature;] because the exertions are with a view to isolation [from all qualities; a condition to which Soul is competent, but Nature is not].

a. The very essence of Nature cannot depart from it [so as to leave it in the state of absolute, solitary isolation contemplated]; because the three Qualities are its very essence, [the departure of which from it would leave nothing behind], and because it would thus prove to be *not* eternal, [whereas, in reality, it *is* eternal]. The isolation (*kaivalya*) of that alone is possible of which the qualities are reflexional, [and not constitutive (see § 58. *a*.)]; and that is Soul.

b. Of what nature is this [Soul]? To this he replies:

Aph. 145.* Since light does not pertain to the unintelligent, light, [which must pertain to something or other, is the essence of the Soul, which, self-manifesting, manifests whatever else is manifest].

a. It is a settled point, that the unintelligent is not light; [it is not self-manifesting]. If Soul, also, were unintelligent [as the Naiyayikas hold it to be, in *substance*; knowledge being, by them, regarded not as its essence or substratum, but as one of its *qualities*], then there would need to be another light for *it*; and, as the simple theory, let Soul itself consist, essentially, of light.

b. And there is Scripture [in support of this view; for example, the two following texts from the *B[ihadaralyaka Upanishad*2]: 'Wherewith shall one distinguish that wherewith one distinguishes all this [world]?' 'Wherewith shall one take cognizance of the cognizer?'

c. [But the Naiyayika may urge,] *let* Soul be unintelligent [in its substance], but have Intelligence as its attribute. *Thereby* it manifests all things; but it is not, essentially, Intelligence. To this he replies:

Aph. 146.* It [Soul,] has not Intelligence as its attribute; because it is without quality.

a. If soul were associated with attributes, it would be [as we hold everything to be, that is associated with attributes,] liable to alteration; and, therefore, there would be no Liberation; [its attributes, or susceptibilities, always keeping it liable to be affected by something or other; or, the absolutely simple being the only unalterable].

b. He declares that there is a contradiction to Scripture in this, [*i.e.*, in the view which he is contending against]:

Aph. 147.* There is no denial [to be allowed] of what is established by Scripture; because the [supposed] evidence of intuition for this [*i.e.*, for the existence or qualities in the Soul,] is confuted [by the Scriptural declaration of the contrary].

a. The text, 'For this Soul is uncompanioned,' &c., would be confuted, if there were any annexation of qualities [to Soul: and the notion of confuting Scripture is not to be entertained for a moment].

b. But the literal meaning [of the aphorism] is this, that the fact, established by Scripture, of its [*i.e.*, soul's,] being devoid of qualities, &c., cannot be denied; because the Scripture itself confutes the [supposed] intuitive perception thereof, *i.e.*, the [supposed] intuitive perception of qualities, &c., [in the soul].

Aph. 148.* [If soul were unintelligent,] it would not be witness [of its own comfort,] in profound [and dreamless] sleep, &c.

a. If soul were unintelligent, then, in deep sleep, &c., it would not be a witness, a knower. But that this is not the case [may be inferred] from the phenomenon, that 'I slept *pleasanty*.' By the expression '&c.' [in the aphorism,] dreaming is included.

b. The Vedant*is* say that 'soul is *one* only'; and so, again, 'For Soul is eternal, omnipresent, changeless, void of blemish:' 'Being one [only], it is divided [into a seeming multitude] by Nature (*shakti*), *i.e.*, Illusion (*maya*), but not through its own essence, [to which there does not belong multiplicity].' In regard to this, he says [as follows]:

Aph. 149.* From the several allotment of birth, &c., a multiplicity of souls [is to be inferred].

a. 'Birth, &c.' By the '&c.,' growth, death, &c., are included.

'From the several allotment' of these, *i.e.*, from their being appointed; [birth to one, death to another, and so on]. 'A multiplicity of souls;' that is to say, souls are many. If soul were one only, then, when *one* is born, *all* must be born, &c.

b. He ponders, as a doubt, the opinion of the others, [*viz.*, of the Vedant*i*s]:

Aph. 150.* [The Vedant*i*s say, that,] there being a difference in its investments, moreover, multiplicity attaches [seemingly,] to the one [Soul]; as is the case with Space, by reason of jars, &c., [which mark out the spaces that they occupy].

a. As Space is one,—[and yet], in consequence of the difference of adjuncts, [as] jars, &c., when a jar is destroyed,. it is [familiarly] said, 'the jar's space is destroyed' [for then there no longer exists a *space marked out by the jar*]; —so, also, on the hypothesis of there being but one Soul, since there is a difference of corporeal limitation, on the destruction thereof, [*i.e.*, of the limitation occasioned by any particular human body], it is merely a way of talking [to say], 'The soul has perished.' [This, indeed, is so far true, that there is really no perishing of Soul; but then it is true,] also on the hypothesis that there are *many* souls. [And it must be true:] otherwise, since Soul is eternal, [without beginning or end, as both parties agree], how could there be the appointment of birth and death?

b. He states [what may serve for] the removal of doubt [as to the point in question]:

Aph. 151.* The investment is different, [according to the Vedant*i*s], but not that to which this belongs; [and the absurd consequences of such an opinion will be seen].

a. 'The investment is different,' [there are diverse bodies of John, Thomas, &c.]; 'that to which this belongs,' *i.e.* that [Soul] to which this investment [of body, in all its multiplicity,] belongs, is *not* different, [but is one only]: such is the meaning. And, [now consider], in consequence of the destruction of one thing, we are not to speak as if there were the destruction of something else; because this [if it were evidence of a thing's being destroyed,] would present itself where it ought not; [the destruction of Devadatta, *e.g.*, presenting itself, as a fact, when we are considering the case of Yajnadatta, who is not, for *that* reason, to be assumed to be dead]: and, on the hypothesis that Soul is one, the [fact that the Vedanta makes an] imputation of in consistent conditions is quite evident; since Bondage and Liberation do not [and cannot,] belong [simultaneously] to *one*. But the conjunction and [simultaneous] non-conjunction of the sky [or space] with smoke, &c., [of which the Vedant*i* may seek to avail himself, as an illustration,] are *not* contradictory; for Conjunction is not pervasion; [whereas, on the other hand, it would be nonsense to speak of Bondage as affecting one portion of a monad, and Liberation as affecting another portion; as a monkey may be in conjunction with a branch of a tree, without being in conjunction with the stem].

b. What may be [proved] by this? To this he replies:

Aph. 152.* Thus, [*i.e.*, by taking the Samkhya view,] there is no imputation of contradictory conditions to [a Soul supposed to be] everywhere present as *one* [infinitely extended monad].

a. 'Thus,' *i.e.*, [if you regard the matter rightly,] according to the manner here set forth, there is no 'imputation,' or attribution, 'of incompatible conditions,' Bondage, Liberation, &c., to a soul 'existing everywhere,' throughout all, as one, [*i.e.*, as a monad].

b. [But, the Vedant*i* may contend,] we *see* the condition of another attributed even to one quite different; as, *e.g.*, Nature's character as an agent [is attributed] to Soul, which is another [than Nature]. To this he replies:

Aph. 153.* Even though there be [imputed to Soul] the possession of the condition of another, this [*i.e.*, that it really possesses such,] is not established by the imputation; because it [Soul,] is *one* [absolutely simple, unqualified entity].

a. [The notion] that Soul is an agent is a mistake; because, that Soul is *not* an agent is true, and the imputation [of agency to Soul] is *not* true, and the combination of the true and the untrue is not real. Neither birth nor death or the like is compatible with Soul; because it is uncompanioned, [*i.e.*, unattended either by qualities or by actions].

b. [But the Vedant*i* may say:] and thus there will be an

opposition to the Scripture. For, according to that, 'Brahma is one without a second:' 'There is nothing here diverse; death after death does he [deluded man,] obtain, who here sees, as it were, a multiplicity.' To this he replies:

Aph. 154.* There is no opposition to the Scriptures [declaratory] of the non-duality [of Soul]; because the reference [in such texts,] is to the *genus*, [or to Soul in general].

a. But there is no opposition [in our Samkhya view of the matter,] to the Scriptures [which speak] of the Oneness of Soul; because those [Scriptural texts] refer to the *genus*. By *genus* we mean sameness, the fact of being of the same nature: and it is to this alone that the texts about the non-duality [of Soul] have reference. It is not the indivisibleness [of Soul,—meaning, by its indivisibleness, the impossibility that there should be more souls than one,—that is meant in such texts]; because there is no motive [for viewing Soul as *thus* indivisible]: such is the meaning.

b. But then, [the Vedant*i* may rejoin,] Bondage and Liberation are just as incompatible in any single soul, on the theory of him who asserts that souls are many, [and that each is at once bound and free]. To this he replies:

Aph. 155.* Of him [*i.e.*, of that soul,] by whom the cause of Bondage is known, there is that condition [of isolation, or entire liberation], by the perception [of the fact, that Nature and soul are distinct, and that he, really, was *not* bound, even when he seemed to be so].

a. By whom is known 'the cause of bondage,' *viz.*, the non-perception that Nature and soul are distinct, of him, 'by the perception' [of it], *i.e.*, by cognizing the distinction, there is 'that condition,' *viz.*, the condition of isolation, [the condition (see § 144) after which the soul aspires. The soul in Bondage which is no real bondage may be typified by Don Quixote, hanging, in the dark, from the ledge of a supposed enormous precipice, and holding on for life, as he thought, from not knowing that his toes were within six inches of the ground].

b. [Well, rejoins the Vedant*i*,] Bondage [as you justly observe,] is dependent on non-perception [of the truth], and is not real. It is a maxim, that non-perception is removed by perception; and, on this showing, we recognize as correct the theory that soul is one, but not that of soul's being multitudinous. To this he replies:

Aph. 156.* No: because the blind do not see, can those who have their eyesight not perceive?

a. What! because a blind man does not see, does also one who has his eyesight not perceive? There are *many* arguments [in support of the view] of those who assert that souls are many, [though *you* do not see them]: such is the meaning.'

b. He declares, for the following reason, also, that souls are many:

Aph. 157.* Vamadeva, as well as others, has been liberated, [if we are to believe the Scriptures; therefore] non-duality is not [asserted, in the same Scriptures, in the Vedantic sense].

a. In the PuraIas, &c., we hear, Vamadeva has been liberated,' '*sh*uka has been liberated,' and so on. If Soul were *one*, since the liberation of all would take place, on the liberation of one, the Scriptural mention of a diversity [of separate and successive liberations] would be self-contradictory.

b. [But the Vedant*i* may rejoin:] on the theory that Souls are many, since the world has been from eternity, and from time to time some one or other is liberated, so, by degrees, *all* having been liberated, there would be a universal void. But, on the theory that Soul is *one*, Liberation is merely the departure of an adjunct, [which, the Vedant*i* flatters himself, does not involve the inconsistency which he objects to the Samkhya], To this he replies:

Aph. 158.* Though it [the world,] has been from eternity, since, up to this day, there has not been [an entire emptying of the world], the future, also, [may be inferentially expected to be] thus [as it has been heretofore].

a. Though the world *has been* from eternity, since, up to this day, we have not seen it become a void, there is no proof [in support] of the view that there will be Liberation [of *all* Souls, so as to leave a void].

b. He states another solution [of the difficulty]:

Aph. 159.* As now [things are, so], everywhere [will they continue to go on: hence there will be] no absolute cutting short [of the course of mundane things].

a. Since souls are [in number,] without end, though Liberation successively take place, there will not be [as a necessary consequence,] a cutting short of the world. As now, so everywhere,—*i.e.*, in time to come, also,—there will be Liberation, but not, therefore, an absolute cutting short [of the world]; since of this the on-flowing is eternal.

b. On the theory, also, that Liberation is the departure of an adjunct [§ 157. *b*.], we should find a universal void; so that the doubt is alike, [in its application to either view]. Just as there might be an end of all things, on the successive liberation of many souls, so, since all adjuncts would cease, when [the fruit of] works [this fruit being in the shape of Soul's association with body, as its adjunct,] came to an end, the world would become void, [on the Vedanta theory, as well as on the Samkhya].

c. Now, [if the Vedant*i* says,] there will not be a void, because adjuncts are [in number,] endless, then it is the same, on the theory that Souls are many. And thus [it has been declared]: 'For this very reason, indeed, though those who are knowing [in regard to the fact that Nature and Soul are different], are continually being liberated, there will not be a void, inasmuch as there is no end of multitudes of souls in the universe.'

d. Pray, [some one may ask,] is Soul [*essentially*] bound? Or free? If [essentially] bound, then, since its essence cannot depart, there is no Liberation; for, if it [the essence,] departed, then it [Soul,] would [cease, with the cessation of its essence, and] not be eternal. If [on the other hand, you reply that it is essentially] free, then meditation and the like [which you prescribe for the attainment of liberation,] are unmeaning. To this he replies:

Aph. 160.* It [Soul,] is altogether free [but seemingly] multiform [or different, in appearance, from a free thing, through a delusive semblance of being bound].

a. It is not bound; nor is it liberated; but it is ever free, [see § 19]. But the destruction of ignorance [as to its actual freedom,] is effected by meditation, &c., [which are, therefore, not unmeaning, as alleged in § 159. *d*.].

b. It has been declared that Soul is a witness. Since it is a witness [some one may object], even when it has attained to discriminating [between Nature and Soul], there is no Liberation; [Soul, on this showing, being not an absolutely simple entity, but something *combined* with the character of a spectator or witness]. To this he replies:

Aph. 161.* It [Soul,] is a witness, through its connexion with sense-organs, [which quit it, on liberation].

a. A sense-organ is an organ of sense. Through its connexion therewith, it [Soul,] is a witness. And where is [its] connexion with sense-organs, [these products of Nature (see § 61)], when discrimination [between Nature and Soul] has taken place?

b. [Well, some one may ask], at all times of *what* nature is Soul? To this he replies:

Aph. 162.* [The nature of Soul is] constant freedom.

a. 'Constant freedom:' that is to say; Soul is, positively, always devoid of the Bondage called Pain [see §§ 1 and 19]; because Pain and the rest are modifcations of Understanding, [which (see § 61) is a modification of Nature, from which Soul is really distinct].

Aph. 163.* And, finally, [the nature of Soul is] indifference [to Pain and Pleasure, alike].

a. By 'indifference' is meant non-agency. The word *iti* [rendered 'finally,'] implies that the exposition of the Nature of Soul is completed.]

b. [Some one may say, the fact of] Soul's being an agent is declared in Scripture. How is this, [if, as you say, it be *not* an agent]? To this he replies:

Aph. 164.* [Soul's *fancy* of] being an agent is, through the influence [of Nature], from the proximity of Intellect, from the proximity of Intellect.

a. [Its] 'being an agent,' *i.e.*, Soul's fancy of being an agent, is 'from the proximity of Intellect,' 'through the influence' of Nature, [(see § 19,) of which Intellect (see § 61) is a modification].

b. The repetition of the expression 'from the proximity of Intellect' is meant to show that we have reached the conclusion: for thus do we see [practised] in the Scriptures, [*e.g.*, where it is said, in the Veda: 'Soul is to be known; it is to be discriminated from Nature: thus it does not come again, it does not come again'].

c. So much, in this Commentary on the illustrious Kapila's Aphorisms declaratory of the Samkhya, for the First Book, that on the [topics or] subject-matter [of the Samkhya system].

3

The Other Systems

The Yoga Philosophy

1. Much has been written and said on the mystic philosophy of ancient nations of the Egyptians, Greeks and Hindus. But I doubt whether it has been rightly understood. The advocates of modern science, some of them base the science of ethics on expediency, others on utility, while there are many to whom moral code is a commandment from a superior to an inferior. Thou shalt commit no murder. Why? The theologian would say-Because that is the commandment of God. The materialists will say-because that is the command of the ruling authority of the state. But why should God and the sovereign issue commands? There is no rational reply. A system of ethics not based on the rational demonstration of the universe is of no practical value. It is only a system of the ethics of individual opinions and individual convenience. It has no solidity and therefore no strength. The aim of human existence is happiness, progress; and all ethics teach how to attain one and achieve the other. The question however remains-What is happiness and what is progress? Those are issues not yet solved in any satisfactory manner in the West by the known systems of ethics. The reason is not far to seek. The modern tendency is to separate ethics from physics or rational demonstration of the universe and thus make it a science resting on nothing but the irregular whims and caprices of individuals and nations.

In India ethics has ever been associated with religion. Religion has ever been an attempt to solve the mystery of nature. Every religion has its philosophical as well as ethical aspect and the latter without the former has in India at least no meaning. If every religion has its physical and ethical side, it has its psychological side as well. There is no possibility of establishing a relation between physics and ethics but through psychology. Psychology enlarges the conclusions of physics and confirms the idea of morality.

The *Yoga* philosophy then is based on the idea that if man wants at all to understand his place in nature and to be happy and progressing he must aim at that physical, psychological and moral development which can enable him to pry into the depths of nature. He must observe, think and act, he must live, love and progress. His development must be simultaneous on all the three planes. The law of correspondence, according to this philosophy, rules supreme in nature and the physical corresponds as much to the mental as both in their turn correspond to the moral. Unless man arrives at this stage of corresponding and simultaneous development on the three planes he is not able to understand the meaning of his existence or existence in general, nor even to grasp the idea of happiness or progress. To that man of high aim whose body, mind and soul act in correspondence the higher, nay, even all, secrets of nature become revealed. He feels within himself as everywhere that Universal Life wherein there is no distinction, no sense of separateness, but therefore all bliss, unity and peace.

Lest I may be misunderstood as subscribing to the doctrine of *Yoga* philosophy except Jainism, I should tell you beforehand that what I am saying here is merely the doctrine of the *Yoga* Philosophy. In my theory in the highest spiritual plane, physical form is not a necessity for the realization of the highest truth. Form is only required in the infant state of development.

The peace of Universal Life then is according to the *Yoga* philosophy the peace of spiritual bliss *Moksh*. The course of nature never ceases, action always compels even the peaceful to act; but the individual being already lost in the All there is nothing unpleasant to disturb. The peace of spiritual development is indescribable and so are its powers indescribably vast. As you go on forgetting yourself, just in the same proportion do spiritual peace and spiritual powers flow towards you. When one consciously suppresses individuality by proper physical, mental moral and spiritual development he becomes part and parcel of the immutable course of nature and never suffers. This fourfold development and spiritual peace have been considered the end of philosophy. In India there have been six such schools of thought. Each starts with a more or less rational demonstration of the universe and ends with a sublime code of ethics. There are first the atomic *Vaisheshika* and the dialectic *Naya* schools seeking mental peace in devotion to the ruler of the universe. Then there are the materialist *Samkhya* and the practical *Yoga* schools teaching mental peace by proper analysis and practical training. Lastly there are the orthodox *Mimamsa* and the Unitarian *Advaita* schools, placing spiritual bliss in strict observance of *Vedic* injunction and in realizing the unity of the Cosmos. It will thus be seen that *Yoga* is a complement of the *Samkhya*.

2. I told you last time when we met that the *Samkhya* philosophy starts with the proposition that the world is full of miseries of three kinds physical, supernatural and corporeal and that these are the results of the properties of matter and not of its correlate intelligence of consciousness, that out of the primordial essence *Prakriti* comes out the whole universe, by reason of the predominance of one or other of the three qualities of *Sativa, Rajas* and *Tams* passivity, activity, all grossness, darkness, ignorance of *Tams*, all pleasure, passivity, knowledge, peace of *Sativa*. The mind is a result of *Rajas* _ and it is *Sativa* alone which by its light illumines it and enables it at times to catch glimpses of the blissful *Purush* ever near to the *Sativa*. As mind or the thinking principal plays an important part in the *Samkhya* and more so in the *Yoga* philosophy, for its chief article is 'Stop the transformation of the thinking principal and you will realize the Self', we will come to a consideration of the mind.

3. With the philosophers of the West, mind and soul are synonyms. The popular definition is-mind is the intellectual power in man. In the East there is a difference of opinion on this subject among the several philosophers. The followers of

the *Naya* philosophy hold that all bodies having a form are impermanent but the mind being formless is permanent; it has special attributes and is likewise subtle; hence it is unable to grasp two objects at the same time. The *Samkhya* philosophy however of which *Yoga* is the complement considers the mind to be a derivative product. Till the *Purush*-soul-is emancipated from *Prakriti* the mind continues in a state of integrity. Its span of duration is limited to a *Mahapralaya*-the great Deluges when it disintegrates to be taken up by *Prakriti*. The seat of the mind has been the subject of an able discussion amongst the ancient philosophers. The followers of the *Purana*s and the *Tantrums* fix it in the forehead near the junction of the two eye-brows. The anatomical description would incline us to look upon the optic thalamus as the center of the mind. The *Vedanta's* hold the mind to be situated in the heart, for they say when an individual thinks of a subject he keeps it next to his heart as in the act of worshipping. There are some philosophers who identify the mind with the soul but Kapila refutes their views. He says: If mind and soul were one and the same, one would say 'I am the mind' instead of 'my mind, my hands'. According to him all experience consists of mental representation, the *Satva* being clouded, obscured or entirely covered over by the nature or property of representation. This is the root of evil. The act of the mind cognizing objects or, technically, taking the shape of objects presented to it is called *Verity* or transformation. It is the *Verity* which being coloured by the presentation imparts the same colour by representation to *Satva* and causes evil, misery, ignorance and the like. All objects are made of three *Gun* or qualities and when the *Verity* or the transformation of the thinking principal sees everywhere nothing but the *Sativa* to the exclusion of the other two, presentation and representation become purely *Satvik* passive and the internal *Sativa* of the cognize realizes itself everywhere and in everything. In the clear mirror of the *Sativa* is reflected the bright and blissful image of the ever present *Purush* who is beyond change, and supreme bliss follows. This state is called *Sativapati* or *Moksa* or *Kevalya*. For every *Purush* who has thus realized itself *Prakriti* has ceased to exist, in other words, has ceased to cause disturbance and misery. The course of nature never ceases but one who receives knowledge remains happy throughout by understanding the truth. The *Samkhya* tries to arrive at this result by a strict mode of life accompanied with analysis and contemplation.

This state of peace besides being conducive to eternal calm and happiness is most favorable to the apprehension of the truths of nature. That intuitive knowledge, which is called *Tarka,* puts the students in possession of almost every kind of knowledge he applies himself to. It is indeed this fact on which the so-called powers of *Yoga* are based.

4. The *Yoga* philosophy subscribes to this *Samkhya* theory in toto. It however appears to hold that *Purush*-Soul-by himself cannot easily acquire that *Satvik* development which leads to knowledge and bliss. A particular kind of *Eashwar* or Supreme God is therefore added for the purposes of contemplation etc. to the twenty-five categories of the *Samkhya*. This circumstance has obtained for *Yoga* the name of *Saishvar Samkhya* or theistic *Samkhya* as the *Samkhya* proper is called *nireashwar Samkhya* or atheistic *Samkhya*.

5. The second and really important improvement on the *Samkhya* consists in the highly practical character of the rules laid down for acquiring eternal bliss and knowledge. The end proposed by the *Yoga* philosophy is *Samadhi* leading to *kaivalya*. *Yoga* and *Samadhi* are convertible terms, either meaning Vritinirodh or suspension of the transformations of the thinking principal.

6. With this introduction we will enter into the details of this philosophy. We have defined *Yoga* to be the suppression of the transformation of the thinking principal. What is the thinking principal and what are its transformations and what results are achieved by the practice of *Yoga*? As to its power it teaches that the powers of electricity and magnetism are but a drop in the ocean compared with those of the soul, when they are fully developed by the practice of *Yoga*. But this is no part of true *Yoga*, although the lower form of *Yoga* does teach, how to develop these powers. The scope of true *Yoga* lies in the realization of the immortal part of man and the keynote of this

self-realization lies in the suppression of the transformation of the thinking principal.

The thinking principal is a comprehensive expression equal to the *Sanskrit* word *Antakaran*, which is divided into four parts-(I *Manas* or mind, the principal which cognizes generally; (ii) *Chit* or individualizing, the idea which fixes itself upon a point and makes the object its own by making it an individual; (iii) *Ahamkar* or egoism, the persuasion which connects the individual with the self; and (iv) *Buddhi* or reasons, the light that determines one way or another. Knowledge or perception is a kind of transformation *Parin am* of the thinking principal into anything which is the subject of external or internal presentation, through one or other of these four.

All knowledge is of the kind of the transformation of the thinking principal. Even the will, which is the very first essential of *Yoga,* is a kind of such transformation. *Yoga* is a complete suppression of the tendency of the thinking principal to transform itself into objects, thoughts etc. It is possible that there should be degrees among these transformations and the higher ones may assist to check the lower ones, but *Yoga* is acquired only when there is complete cessation of the one or the other. It should distinctly be borne in mind that the thinking principal in this philosophy is not the soul who is the source of all consciousness and knowledge. The suppression of the transformations of the thinking principal does not therefore mean that the yogi-the practitioner of the *Yoga*-is enjoined to become all, which is certainly impossible.

The thinking principal has three-property passivity, activity and grossness. When the action of the last two is checked the mind stands steady like the jet of a lamp in a place protected from the least breeze. When all the transformation of the thinking principal are suppressed there remains only the never changing eternal soul-the *Purush*-in the perfect *Sata* passivity. Otherwise when the thinking principal transforms itself into objective and subjective phenomena the *Purush* is for the time obscured by it or which is the same thing assimilated into it. It is only when the state of *Yoga* is reached that the consciousness becomes quite pure and ready to receive all knowledge and all impressions from any source whatever. If this state is to be acquired by suppressing the transformations of the thinking principal, let us see what these transformations are.

7. In *Yoga* philosophy the thinking principal is modified in five ways. First when there comes to it the right knowledge, second when there comes to it false knowledge, third when it is simply put into complex imagination or fancy, fourth when we are sleeping and fifth when we are exercising the faculty of memory. Let us examine each condition. The theory as to how the external world is cognized is a complicated one, but in order to explain it in the simplest way it will do to say [the following]. When organs of sense are put in contact with external objects they are put in to a state of vibration and cause a similar vibration on he mind-substance. This charge in the mind-substance is called direct cognition. It is only one kind of right knowledge. The mind is also transformed when it infers or draws conclusions and also when it receives knowledge from words of authority-trust worthy authority. These three kinds of knowledge are collectively known as right knowledge. When the mind cognizes in any of the three ways there is a corresponding motion or change produced in it.

That is one way in which mind becomes subject to transformation. The second way in which it is modified is false knowledge. This is when a false conception is entertained of a thing whose real form does not correspond to that conception, for instance, when a mother of pearl is mistaken for silver or a post mistaken for a man. The third way in which the mind is modified is by having fancied notions, *i.e.* notions called into being by mere words having nothing to answer to them in reality. The fourth way in which the mind is transformed is sleep and the fifth way is the exercise of memory, *i.e.* by recollection impressions of past experience. It may be remarked that of these five kinds of transformations of the mind, right knowledge, false knowledge and fancy belong to the waking state. When any of these becomes perceptible in sleep it is dream. Sleep itself has no cognition. Memory may be (may depend on?) any of them.

8. Now the suppression of these transformations is the *Yoga,* which leads to the realization of the Self. What are the means of suppressing them? The author of the *Yoga Sutras* says that complete suppression of the transformation of the mind is secured only by sustained application and non-attachment. Application is of course steady sustained effort to reach that state and non-attachment is the consciousness of having mastered every desire for any object. And further rules are given for the purpose of rising to that high state of self-knowledge.

9. But in the meantime I will draw your attention to the fact that some scholars like Monier Williams and others have thought that this system of *Yoga* is nothing but a mere contrivance of getting rid of all thought and that it is a strange compound of mental and bodily exercises, consisting in unnatural restraint, forced and painful postures, twisting and contortions of the limbs, suppression of the breath and utter absence of mind. In the opinion of such scholars it is not possible that a man should actually know any thing transcending his sensual perception unless it is told to him by some supposed authority. In their opinion the power of intuition cannot be developed to such an extent as to become actual knowledge without any possibility of error and we shall always be doomed to depend upon hearsay and opinions. To them extra-ordinary powers of the soul are mere dreams. The author of the 'Modern Science and Modern Thought' says: "Almost the entire world of the supernatural fades away of itself with an extension of our knowledge of the laws of nature, as surely as the mists melt from the valley before the rays of the morning sun. We have seen how throughout the wide domains of space, time and matter, law uniform, universal and inexorable reigns supreme, and there is absolutely no room for the interference of any outside personal agency to suspend its agency (Hindus have never said so). The last remnant of supernaturalism therefore, apart from Christian Miracles which we shall presently consider, has sunk into that doubtful and shady borderland of ghosts, spiritualism and mesmerism, where vision and fact and partly real partly imaginary effects of abnormal nervous conditions

are mixed up in a nebulous haze with a large dose of imposture and credulity." These are the words of a famous English writer. Let us hear then what the neigbour of the John Bull says in regard to the claim of the modern scientist. Dr. Heinrich Hensoldt of Germany says: "Apart from the material progress or mere outward development which the Hindoos had already attained in times which we are apt to call pre-historic as evinced by the splendor of their buildings and the luxuries and refinements of their civilization in general, it would seem as if this greatest and most subtle of Aryan races had developed an inner life even more strange and wonderful. Let those who are imbued with the prevalent modern conceit that we Westerners have reached the highest pinnacle of intellectual culture, go to India. Let them go to the land of mystery, which was ancient, when the Great Alexander crossed the Indus with his warriors, ancient, when Abraham roamed the plains of Chaldea with his cattle, ancient when the first pyramid was built, and if after a careful study of Hindoo life, religion and philosophy, the inquirer is still of opinion that the palm of intellectual advancement belongs to the Western world-let him lose no time in having his own cranium examined by a competent physician." These are the words of Dr. Hensoldt.

10. Without caring much what the foreigners have to say in reference to the religions and philosophies of India we will come to our own subject. We have said before that *Yoga* is the suppression of the manifestations of the mind. The source of the positive power therefore lies in the soul. In the very wording of the definition of *Yoga* is involved the supposition of the existence of a power which can control and suppress the manifestations of the mind. This power is the power of the soul-otherwise familiar to us as freedom of the will. So long as the soul is subject to the mind it is tossed this way or that in obedience to the mental changes. Instead of the soul being tossed by the mental changes, the mind should vibrate in obedience to the soul-vibrations. When once the soul becomes the master of the mind, it can produce any manifestations it likes. The ancient Chaldeons and the modern monks of India, Japan and China teach the same doctrine. It was by the aid

of this *Yoga* science that the ancients made many discoveries in chemistry and medicine.

11. We will now come to our point. The suppression of all mental modification produces the state called *Yoga* or *Samadhi*. This *Samadhi* is of two kinds *Svikalp* and *nirvikalpa*. The first is that in which the mind is at rest only for the time, the other is that in which through supreme universal non-attachment it is centered in (passivity) *Satva* and realizes *Satva* everywhere for all time. The mind being as it were annihilated *Purush*-the soul-alone shines in native bliss. This is called *Kaivalya*. This is the end view. This is the summum onum, the end and aim of philosophy. Between this end and the first stage of mental suppression there are several stage. The author of the *Yoga* aphorisms mentions eight stage; they are *Yam, Niyam, Aasan, Pran am, Pratyahar, Dharn a, Dhyan, Smaddhi* This leads us to the practical part of *Yoga*.

12. (a)The first stage is *Yam*. What a student of *Yoga* is required to do in the first stage is forbearance or control over mind, body and speech and it consists in abstaining from killing, falsehood, theft, incontinence and greediness. (i) The first of these is killing-*Hinsa* in *Sanskrit*. It is difficult to give the full meaning of this word *Hinsa*. It means wishing evil to any being by word, act or thought and abstinence of this kind of killing is the first requirement of a student of *Yoga*. It obviously implies abstinence from animal food in as much as it is never procurable without direct or indirect *Hinsa* of some kind. Not with standing the sanction given by the *Veda*s to the system of sacrificing animals to gods, the Hindu scriptures are very strong on this point when they treat of the practical part of the *Yoga* philosophy. Manu, the great law-maker of the Hindus, says:

Anuyanta vishsita nihanta kryavikryee

Sanskatee chopharta cha khadkshchaitee ghatak

[One who indirectly gives permission to kill animals, one who separates the several parts of an animal after it is killed, one who actually kills the animal, one who sells meat, one who cooks meat, one who serves meat at the table and one who eats it are all considered killers of the animal.]

Akritva pran·inan hinsan mansan notpadyatai kachit

Na cha pran·ivdhat svarg tsmanmansan vivrjyait

[You cannot get meat unless an animal is killed, killing of animals can never lead to a higher state, therefore abstain from meat altogether.] The avoidance from animal food from another point of view is strongly recommended, as it always leads to the complete obscuration and even annihilation of intuition and spirituality. It is to secure this condition of being ever with nature and never against it or, in other words, being in love with nature that all other restrictions are prescribed. (ii) The next requirement is abstaining from falsehood, *i.e.* from telling what we do not know or believe to be the exact state of things. (iii) The third thing to be avoided includes, besides actual illegal appropriation, even the thought for any such gain. (iv) So also does incontinence, the fourth danger in the path of success, include, besides physical enjoyment, even talking to, looking at or thinking of the other sex, with lustful intention. And here we come to the very important point of view of celibacy. We know that even doctors of eminence talk about the dictates of nature-as if animosity and brutality are natural parts of man. They may talk about sexual needs, imperious necessities, and uncontrollable passion. But when we come to the actual state of facts, we will realize the truth. We know that the trainer of a pugilist denies his man all sexual indulgence whatever, the trainer on a boat's crew would abandon all hope of victory if he knew that his men visited women even once a week. Indeed so jealous is he that he will not permit his wards even to talk much with the other sex, lest some erotic fancy should affect the condition of their nerves. An eminent doctor of the United States says: "All eminent physiologists who have written on this point agree that the most precious atoms of the blood enter into the composition of the semen. A healthy man may occasionally discharge his seed with impunity, but if he chooses-with reference as in the pedestrian, boat-racer, prize-fighter or explorer or with reference to great intellectual and moral work as in the apostle Paul, Sir Isaac Newton and a thousand other instances-to refrain from sexual pleasure, nature well knows what to do with those precious atoms. She finds use for them in building up a keener brain and more vital and enduring nerves and muscles." The chief monk of my community Muni Atmaramji was once asked by a Hindu gentleman, how it was that in running contrary to the course of nature-*i.e.* not obeying the urgent demands of natural instincts in such nature-he could build up his constitution which could well defy the attacks of an athlete or a stalwart. The monk in reply simply recited a verse:

Sinho balee dvirdashookrmansjeevee
Sanvtsrain· Ratimaiti kilaekvaram
Paravat kharshilakan·matrjeevee
Kamee bhavtynudinan vad kotr haitu

[The lion eats the flesh of elephants and hogs and is the strongest of all animals, still he enjoys sexual intercourse only once in a year, while doves and pigeons that live on dirt and sorts of refuse are lustful every day.] (v) The last of the five forbearances is greediness. It consists not only in coveting more that necessary but also in keeping in possession anything beyond the very necessaries of life. Some practitioners are known to carry this requirement to the extent of even not accepting anything whatever from others. We thus finish the list of five kinds of forbearances; that is the first stage through which a student of *Yoga* has to pass.

(b) The second stage is *Niyam*, *i.e.* observances. They are also five, purity, contentment, austerity, study and resignation to *Eashwar*,-the Lord. The five kinds of forbearances, which we mentioned before, were negative injunctions, the five kinds of observances, which we are now describing, are positive commands. (i) The first in purity, *i.e.* purity bodily and mental which latter consists in universal love and equanimity. (ii) The second is contentment-being satisfied with one's lot. (iii) The third is austerities, *i.e.* fasts, penances, observances etc. mentioned in the Hindu Dharma Shastras. (iv) Study-the fourth-is the repetition of the sacred mystic word OM or any other holy incantation. (v) Resignation to *Eashwar* the fifth observance-means that the practitioner should so abandon himself to the will of the Supreme that he must move about only to fulfil his benign wish, not to accomplish this or that result. He must bear all good, bad or indifferent, simple as an act of his grace in carrying which he only pleases him. The five kinds of forbearances and the five kinds of observances make ten.

13. (a) (i) The first forbearance was abstinence from killing. What is its result? When one has acquired that confirmed frame of mind-the positive feeling of universal love for all living creatures, even natural antipathy is held in abeyance in his presence; needless to add that no one harms or injures him. All beings, men, animals, birds approach him without reserve. In an extended description of the religious rites, monastic life and superstitions of the Siamese dela loubete cites among other things the wonderful power over wild beasts possessed by the Talapoin (the monks or the holy men of Buddha whose first injunction was protection of all living beings). "The Talapoin of Siam", he says, "will pass whole weeks in the dense woods under a small awning of branches and palm leaves and never make a fire in the night to scare away the wild beasts, as all other people do who travel through the woods of this country. The people consider it a miracle that no Talapoin is ever devoured. The tigers, elephants and rhinoceroses-with which the neighborhood abounds-respect him and travelers placed in secure ambuscade have often seen these wild beasts lick the hands and feet of the sleeping Talapoin." The Jaina history also testifies to the same fact. Mahavira-The twenty-fourth prophet of the Jainas who lived 600 years before Christ-is reported to have attracted, by the sweetness of his musical sermons in parks, wild beasts and animals who stood before him in perfect peace and harmony. Even in the present times no wild beast is known to have devoured a Jaina in India whose first principal is the protection of life-even of the tiniest insect. Strange to say that the Western powers and nations attempt to restore peace and harmony among people by the sharpest swords, huge man-killing machines and animal-food.

(ii) The second forbearance of the five we mentioned before is truthfulness. What is the result? When entire and unswerving truthfulness is fully established, all thoughts and words become immediately effective. What others get by act such as sacrifice to deities etc. He gets by mere thought or word. Emperor Marcus Aurelius says: "He who acts unjustly acts impiously, for since the universal nature has made rational animals for the sake of one another, to help one another according to their deserts, but in no way to injure one ancther, he who transgresses his will is clearly guilty of impiety towards the

highest divinity. And he too that lies is guilty of impiety to the same divinity, from the universal nature of all things that are; and all things that are have a relation to all things that come into existence. And future this universal nature is named Truth and is the prime cause of all things that are true. He then who lies intentionally is guilty of impiety in as much as he acts impiously by deceiving and he also who lies unintentionally in as much as he is at variance with the universal nature, and in as much as he disturbs the order by fighting against the nature of the world; for he fights against it, who is moved of himself to that which is contrary to truth, for he has revived powers from nature, through the neglect of which he is not able now to distinguish falsehood from truth. And indeed he who pursues pleasure as good and avoids pain as evil is guilty of impiety."

What is true of individuals is true of nations. We know that Spain, Greece and Turkey are dishonored in the commercial world. His riches killed Spain. The gold which came pouring into Spain from her vanquished colonies in South America depraved the people, and rendered them indolent and lazy. Now a day, Spaniards would blush to work. He will not blush to beg. The same has been the case with Greece also. She has repudiated her debts for many years. Like Turkey she has nothing to pay. All the works of industry in those countries, are done by foreigners.

Much better things might have been hoped from Pennsylvania and other American states, which repudiated their debts many years ago. They were rich states and the money borrowed from abroad made them richer, by opening roads and constructing canals for the benefit and privation" it was he who was the congress at Washington which he afterwards published "The Americans", he said, "who boast to have improved the institutions of the old world have at least equaled its crimes. A great nation after trampling under foot all earthly tyranny has been guilty of a fraud as enormous as ever disgraced the worst king of the most degraded nation of Europe."

But the state of Illinois acted nobly though it was poor. It had borrowed money like Pennsylvania, for the purpose of carrying out internal improvements. When the inhabitants of rich Pennsylvania set the example of repudiating their footsteps. As every householder had a vote it was easy, if they were dishonest, to repudiate their debts.

A Convention met at Springfield and the repudiation ordinance was offered to the meeting. It was about to be adopted, when an honest man stopped it. Stephen A. Douglas was being sick at his hotel, when he desired to be taken to the Convention. He was carried on a mattress, for he was too ill to walk. Lying on his back he wrote the following resolution, which he offered as a substitute for the repudiation ordinance:

"Resolved that Illinois will be honest although she never pays a cent."

The resolution touched the honest sentiment of every member of the Convention. It was adopted with enthusiasm. It dealt a deathblow to the system of repudiation. The canal bonds immediately rose, capital and emigration flowed into the state and Illinois is now one of the most prosperous states. She has more miles of railway than any of other states. Her broad prairies are one great grain field and are dotted about with hundreds of thousands of peaceful happy homes. This is what truthfulness does. It this is true in the science of nations how much more is it true in the highest known science-the *Yoga*?

(iii) [The last time we left our subject with the result, which can be, worked out from the second kind of forbearance the truthfulness. We will proceed with the rest of them.] The third kind of forbearance is abstinence from self-love and desire of misappropriation. To him who has given up this, all jewels and wealth stumble at his feet even without seeking them.

(iv) The fourth kind of forbearance is continence. On this

subject we dwelt at some length the last time. The point settled in this *Yoga* philosophy is that it is a physiological law that the creative essence in man is closely connected with the intellect and spirituality. Waste of this spiritual element means waste of bodily and mental powers. Preservation of this elements means the acquisition of (?) powers of the brain and body. No *Yoga* is ever reported successful without the observance of this rule as an essential preliminary.

(v) The fifth kind of forbearance is abstinence from greediness. The *Yoga* philosophy teaches that when desire is destroyed, when in fact even the last and subtle but unconquerable desire for life too is given up, there arises knowledge of the why and wherefore of existence.

(b) We mentioned last time five forms of observances. They are purity, bodily and mental, contentment, austerities, study and resignation to *Eashwar*. (i) It is needless to say that mental purity leads to passivity, pleasantness, fix attention, subjugation of the senses and fitness for communion with soul. (ii) The second observance is contentment. Superlative happiness is the result of contentment. (iii) As for the austerities, the *Yoga* philosophy claims that miraculous powers of the body and the senses arise therefrom; the inner sense becomes more developed in proportion to the mortification of the flesh and various methods more or less severe are practiced in all religions. Miraculous powers known as second sight, levitation etc. are the result of austerities. Even some ignorant classes of India are known to possess these powers. They are accounted to flow on account of austerities practiced in past incarnation though in ignorance of the laws of such powers. Although these are the sign of the real *Yoga* power, they are not the true end of *Yoga*. (iv) Study-the fourth observance-is claimed to lead to communion with the higher and subtler forces of nature. The constant silent and devoted repetition of certain formulas is said to be efficacious in establishing a sort of communion with the higher powers of nature. (v) And the resignation to *Eashwar* leads to the accomplishment of that final state of quietude, the *Samadhi* We have then finished the first two stages through which a practitioner to *Yoga* has to pass.

14. The third stage is posture. Various modes of keeping the body in position at the time of perfuming *Yoga* are given in different books. The general and most convenient definition of posture is that it should be perfectly steady and should cause no painful sensation. There is a class of yogis in India who hold that the breath in the body is a part of the universal breath and that the health of mind and body accompanied by spiritual bliss and knowledge will follow on controlling the individual breath in such a manner as to attune it to the cosmic breath. Their methods are more physical than mental. They give much attention to the different postures of the body to be assumed while practicing the *Yoga*. These postures are said to be 84 in number and each has its peculiar influence in the body and the mind. By various kinds of postures and modes of controlling the breath the yogis get over almost all kinds of diseases. Of these postures four are considered the best for *Yoga* practice. The first is *Swastic* posture. In that posture you have to sit with the body perfectly straight placing the right foot in the cavity between the left thigh and calf and the left foot in the cavity between right thigh and the calf. The second is the *Sidh* posture, the third is the *Padm* posture and the fourth is the *Bhadra* posture. As none of us is ready and willing to pass through all the difficult stages of the *Yoga* it is needless to describe these postures. Suffice it to say that the *Hathyogi* having mastered one of the postures commences the actual practice of *Yoga*. *Hatha Yoga* Pradipika-the text book of these yogis says: "One who abstains totally form sexual intercourse, keeps temperate habits and remain free from worldliness becomes a yogi after a full twelve month's practice. By temperance in eating is meant the eating only three fourths of what is actually required. The food also should consist of substantial liquids and solid. Bitter, acid, pungent, salty and hot things as well as green vegetables, oil, intoxicating drugs, animal food of every description, curds (?) etc. are to be strictly avoided. Wheat, rice, barley, milk, butter, sugar, honey, dry ginger, oats and natural water are most agreeable. In the beginning avoid fire, sexual intercourse and extreme exertion. Young, old, decrepit or sick may all obtain success by study practice; none succeeds who lacks in practice; mere reading of

Yoga books or talking on the subject can never conduce to success."

15. The fourth stage through which a student has to pass is Pran*ayam* the control of the expiration and inspiration of the breath. It does not mean that there ought to be an unnatural flow or control of the breath; it means rather that the breath should be controlled or allowed to flow in accordance with the result to be attained. There are three kinds of *Pranayam*. When the breath is expired or held out it is called *Raichak* the first *Pranayam*. When it is drawn in it is called *Poorak* the second *Pranayam*. When it is suspended all at once it is called *kumbhaka* the third *Pranayam*. These three are again regulated by time. Works on *Yoga* say that three kind of *Pranayam* are often to be combined in one single act and their number should be slowly and slowly carried to eighty every time one sits for practice. There are other works, which say that the number must be sufficient to enable the student to mark the first Udghat and follow it afterwards. By *Udghat* they mean the rising of the breath form the navel and its striking at the roof of the palate. *Pranayam* has its chief object the mixing of *Pran* the upper breath and *Apan* the lower breath and rising them upwards by degrees and stages till they subside in called *Kudlini*. It is this force which is the source of all occult powers. The general practice is to begin with *Raichak* followed by *Poorak* by the same nostril, whence the control is begun over again with *Poorak* and onward. This is called one *Pranayam*. The *Hatha Yoga Pradipika* says on this subject as follows:

"Having mastered some one posture and observing the rules of etc. the yogi may begin the study of regulating the breath. Disturbance of mind follows disturbance of breath and mind remains calm when the breath is calm; hence in order to attain fixing of mind the breath should be controlled. So long as the *Nadee* the vehicles of *Pran* Are obstructed by abnormal humors, there is no possibility of the *Pran* Running the middle course *Sushumana* and of accomplishing the *Unmani mudra*. Hence *Pranayam* should be practiced in the first instance for the clearance of these humors. The *Pranayam* for this purpose is as follow: Having assumed the *Parmesan* posture the yogi should inhale at the left nostril and, having retained the breath for the time he easily can, should let it off at the opposite nostril, and repeat the same process beginning with the nostril where he exhales. This will make one *Pran ayam*. These should be practiced four times in twenty-four hours, in the morning, at noon, in the evening, at midnight, and should be carried to eighty each time. The process in its lowest stage will produce perspiration, in its middle stage tremor, and its highest stage levitation.

The student may rub off his body with the perspiration, for this will make his body strong and light. In the beginning of the practice being mastered no such rule is necessary. The breath should be mastered slowly and by degrees, just as are trained tigers, bears and other wild beasts, for otherwise the rash student is sure to come to grief. Proper *Pran ayam* destroys all diseases, improper one produces them. When the humors of the *Nar* ee are cleared the body becomes light and beautiful and digestion becomes strong, health ensues, the retention of breath is done without effort and the *Nad* (sound) within becomes audible."

The opinion of the *Yoga* author is that by this practice of *Parmesan* the outer covering of the soul-the result of *karma*- is removed and the real nature of the soul is realized once and for ever.

16. This leads us to the fifth stage through which the practitioner has to pass. By the practice of *Pran ayam* the mind becomes fit for being quite absorbed in the subject thought of. It is *Parmesan* the mind becomes fit for being quite absorbed in the subject thought of. It is *Pran ayam*, which leads the way to this state, which is the fifth stage. It is *Pran ayam* (abstraction)-imitating by the senses, the thinking principal by withdrawing themselves from their objects. It consists in the sense becoming entirely assimilated to or controlled by the mind. They must be drawn away from their objects and fixed upon the mind and assimilated to it, so that by preventing the transformation of the thinking principal, the senses also will follow it and will be immediately controlled. Not only that, but they will be ever ready to contribute collectively towards the

absorbing meditation of any given thing at any moment and even always.

17. Passing through these five stages, *Yam, Niyam, Aasan, Parmesan* and *Pratyahar* the *yogi* purifies the inner self by avoiding distraction. We then come to the sixth stage *Dharan* a or contemplation. It is the fixing of the mind on something, external or internal. If internal it may be the tip of the tongue or the nose or any convenient spot. If external it may be any suitable image of the deity, or a picture or any similar object. Of course it is necessary to bear in mind that any such thing contemplated upon externally or internally should be strictly associated with nothing but holiness and purity. The mind should be able to picture in itself the object even in its absence in all vividness and at an instant's notice.

18. The next stage is *Dhyan* or absorption, *i.e.* the entire fixing of the mind on the object thought of to the extent of making it one with it. In fact the mind should at the time be conscious of itself and the object.

19. Proceed a step further and we come to the eighth stage, the *Samadhi*. The absorption is to be carried to the extent of forgetting the act and of becoming the thing thought of. This state of *Samadhi* implies two distinct states of consciousness unified in one. The first-that is trance proper-is the forgetting of all idea of the act and the second-the more important factor-is the becoming the object thought of. Mere passive trance is a dangerous practice as it leads to the madness of irresponsible medium-ship. It is therefore necessary to lay stress upon the second part of the connotation of the term *Samadhi*.

20. The three stages, contemplation, absorption and trance are in fact stages of contemplation; for the thing thought upon, the thinker and the instrument together with other things, which are attempted to be excluded are all present in the first, *i.e.* contemplation, all except the last, *i.e.* two, are present in the second and nothing but the thing is present in the third. This trance *Samadhi* however is not complete *Yoga*, for it is only *Svikalpak* or conscious *Samadhi*, having something to rest upon.

Sanyam is the technical name for these three inseparable processes taken collectively. When the three are successively practiced with respect to the one and the same object at any one time it is called *Sanyam*. But it is practiced by stages. One cannot pass all at once to the highest kind of *Sanyam* any more than one can think of something without first knowing it. For example, when *Sanyam* is practiced with respect to a mental image, the process will tend from contemplating upon the gross to [that upon] the subtle. The image may be thought of in all parts, then without the decoration, then without limbs, then without any special identity and lastly as not apart from self.

21. We have thus finished the eight stages. The first five of them are only the preliminaries to the *Yoga,* which really consists in the last three. The first five accessories are called the external means of *Yoga*. The last three are internal. Even the *Sanyam* as the last three are so called collectively) is merely preparatory for the final end, the unconscious *Samadhi* for in *Sanyam* there is something to depend upon whereas in real *Samadhi* there is nothing to depend upon.

The question therefore naturally arises-what does the mind transform itself into in that state of unconscious *Samadhi*? The transformed state in that *Samadhi* is known in *Sanskrit* as *nirodha*, *i.e.* interception of all transformations, thoughts or distractions-of course not ordinary distractions but the distraction, which is still there in the form of conscious *Samadhi*, conscious *Samadhi*, is a distraction, no doubt, for there is yet something which the mind entirely transforms itself into.

The moment the mind begins to pass from one state into the other, two distinct processes begin *viz.* the slow but sure going out of the impressions that distract and the equally gradual but certain rise of the impressions that intercept. When the intercepting impressions gain complete supremacy, the moment of interception is achieved and the mind transforms itself into this intercepting moment so to speak. It is in the interval of this change that the mind may drop and fall into what is called *Ley* or a state of passive dullness leading to all the miseries of irresponsible mediumship.

Hence this passage from the conscious to the so-called unconscious is a very difficult and critical process. This *Samadhi* is called *Nirodhparin·am* or the transformation of the mind into interceptions. It is called the *Dharmparin·am* or the transformation of the thing's property. The intercepting impressions must rise so often as to become a habit, for then alone their flow will become deep and steady and lead to the highest *Samadhi*. The mind is as it were quite annihilated, for no transformation exists. The permanence of this state is all that is desired.

So this trance-transformation is the setting and rising of distractions and concentration respectively-distractions, *i.e.*, of the mind which draw it off from unconscious *Samadhi i.e.* concentration. Interceptions being repeated gain a certain firmness and ripen into unconscious *Samadhi*. Hence when this stage is reached the mere negative condition becomes as it were positive and there arises concentration on nothing, to use a paradoxical phrase.

The moment when the mind arrives at this stage in its transformations is called *Avasthaparin·*am. What is the state of mind of the moment of complete unconscious *Samadhi*? The mind is conscious of nothing except the respective repression and revival of certain impressions, *viz.* distractions and interceptions, both welded in one act of supreme consciousness. This is called *Avasthaparin·am* or transformation as to condition. The mind has its property first transformed. Then this property is joined to a certain moment of time, then the first transformation becomes perfectly ripe and indicates the real condition of the mind. Then it is easy to see that transformation though essentially one is for the sake of explanation and analysis described as threefold.

23. In the *Yoga* philosophy the theory of transformation of the mind is extended to all objects, for there is nothing which is not compounded of one or more or all of the three properties (passivity, activity and grossness) which are ever in a state of transformation. When the very property of a thing is altered it is called property-transformation or *Dharmprin·*am. When afterwards the thing with its altered property becomes manifest

in relation to some time, past, present or future, it is called its (rather its property's) character transformation or *Lakshanparin·am*, for without the limitation of time it is difficult to characterize or define the nature of any conceivable entity. When after this the particular property thus defined ripens into maturity or decay, it is called its condition-transformation or *Avasthaparin·am*. Thus the whole universe consists of nothing but certain objects and their properties which later by their transformation produce all variety. Thus this philosophy puts forth an explanation of the phenomenal universe in accordance with the doctrine of the Samkhya

Let us now see how the *Yoga* philosophy explains what the object or the substratum of those properties is. The doctrine *'Ex nihilo Nihau fit'* is carried out to its full extent by this school and therefore it is held that anything can never manifest itself in any other thing unless it previously existed there. This manifestation has reference only to the properties of things and it cannot be said what will come out of what. In fact every thing is producible for everything, for everything potentially exists in the root of all, the *Prakriti*. All this however takes place in relation to the form in which a thing manifests itself, and this form is none other than the unique combination of the three original properties. The properties can never exist but in relation to some substratum which in its turn can never become cognizable but through the properties.

The properties, which have once manifested themselves and passed into oblivion are called tranquil, for they have played their part and are still there to become actively manifest some other day. Those that are seen at any moment are called active, whereas those not yet manifest are consigned to the realm of possibility or the indescribable. In other words these possible manifestations are as yet latent.

Thus the object or the substratum of properties is that which is correlated to the properties in one or the other of the three states. In the opinion of the *Yoga* philosophers therefore whatever form anything manifests itself as the phenomena is nothing mare than a mere succession of properties in one or other of the three conditions and the universe with all its phenomena is nothing more than an incessant and immediate succession of states of properties.

24. We have digressed from our discussion of the last three stages of *Yoga* to a discussion of the 'substratum and its transformations' theory of the *Yoga* philosophy. But in doing so I had a purpose. The *Yoga* philosophy claims that by performing *Sanyam* on the transformations the past and future of their substratum is at once revealed to the mind.

25. There is another result claimed by the *Yogists* to follow from *Sanyam* and based on a theory to which we are now coming. Every school of philosophy has its own theory about the relation between word and meaning but it would be sufficient here to observe that the *Yoga* philosophy accepts what is generally known as the *Sphote* doctrine. *Sphote* is a something indescribable which eternally exists apart from the letters forming any word and is yet inseparably connected with it, for it reveals itself on the utterance of that word. In like manner the meaning of a sentence is also revealed, so to speak, from the collective sense of the words used. So then, the eternal sense of a word is always different from the letters making that word; and the knowledge which in its turn is conveyed to our mind is equally apart from these two. The sense of words is generally classified under four heads: objects, properties, actions and abstractions; and the impressions into which our mind transforms itself at the moment of cognizing is the knowledge produce. In ordinary intercourse it so happens that the letters, the sense and the knowledge all are so confused together as not to be separable from one another. Thus letters *i.e.* sounds, being confused with sense and knowledge, convey no precise meaning if they happen to be beyond our previous acquaintance. The fact however is that every meaning is eternally existent and is as eternally connected with particular sounds and therefore conveys or reveals the same sense where ever it is uttered. Acting on this theory and performing *Sanyam* on the three, *i.e.* sounds, sense and knowledge, separately the yogi comprehends the sense of all sounds uttered by any sentient being in nature. Even so can the music of nature be heard and the joyous *Nad* within be cognized and understood.

26. *Sanyam* is also claimed to produce knowledge of former births. In all the philosophies based on the *Vedas*, as well as in Buddhism and Jainism, transmigration of the soul-re-incarnation from one body to another-is the one doctrine which runs parallel in all of them. As to the grounds on which it is based, we will fully discuss them when we come to Jainism. In this place however we have simply to refer to the *Yoga* doctrine that by performing Sanyam, which is the same thing as complete mental presentation, on the impressions inherent in the mind from time immemorial, there arises knowledge of previous incarnations.

27. *Yoga* also claims that *Sanyam* leads to the power of mind reading. When a yogi performs Sanyam with reference to any sign as the complexion, the voice or any such thing, he at once understands the state of the mind of which these are the sure indices. Anybody's mind can thus be easily comprehended by the yogi, *i.e.*, he understands the state of the mind. In order to understand the subject occupying the mind of that person, he has of course to perform *Sanyam* on that subject.

28. This philosophy also claims that by performing. *Sanyam* in a certain way you can even cause the body to disappear. The theory on this point is this. When light, the property of *Sativa*, emanates from our body and becomes united with the organ of sight which again is a reservoir of similar light, visual perception follows. Following this theory {when} the yogi performs *Sanyam* on the form of his body, *i.e.* the property that endows visibility to his body, he disserves the connection between the light from his body and the eye of the cognizer and thus follows the disappearance of the body. The yogi in fact centers all this visibility in his thinking principal and prevents the perception of his body. The same holds true of the other organs of his sense and hence of sounds, sensations (touch sensations?) etc.

29. There is a *Sanyam* on *karma* also *Karma* of course means past actions and they are divided by the *Yoga* philosophy into two divisions-active and dormant. That *karma)* which produces its result speedily and is actually on the way to bear

fruit is called active, whereas that which is only in a latent condition of potency is called dormant. By performing *Sanyam* on these two classes of *Karma* the *yogi* knows the time of the cessation of his life. He knows at once which will produce what fruit and therefore at once sees the condition of his death.

The same knowledge also arises from portents in the case of a yogi. Portents are corporeal, celestial or physical. The corporeal are such as the inaudibility of the Pran in the stomach on closing the ears. The celestials are such as the sight of things generally regarded invisible as heaven etc. The physical consists in seeing extra-ordinary or frightful beings etc. These and similar portents such as dreams, the chance hearing of certain words etc. indicate, to use a common expression, which way the wind blows. But none but yogis can make use of any such portents, for it is only they who can precisely interpret them.

30. The *Sanyam* has also its variety of effects by being practiced things. Of course, by *Sanyam* we mean three stages; contemplation, absorption and trance. By performing *Sanyam* with reference to sympathy, compassion and complacency each of these feelings becomes so strong as to produce the desired result at any moment. In fact he finds no difficulty in enlisting the good will and friendship of any one at any moment.

31. By performing *Sanyam* on the powers of elephant or any animal the yogi acquires those powers. By contemplation on the inner light of the *Sanyam* is acquired the knowledge of subtle things such as invisible atoms, obscure things such as hidden treasures and mines and things which are unapproachably remote. By contemplation on the sun, the knowledge of the space intercepted between the earth and the sun is acquired. By contemplation on the moon the knowledge of the starry region is acquired. By contemplation on the pole star is produced the knowledge of the relative motions and positions of the stars and planets. Such are the powers which the contemplation on the external world brings to the yogi.

32. We now come to powers, which he obtains by contemplation on the parts of his body. In the *Yoga* philosophy the theory is that there are *padma* or plexuses formed by

nerves and ganglia at different places in the body. The are generally believed to be seven in number. The most important of these, so far as the arrangement of the nerves of the body is concerned, is the *Nabhichakra* or the navel circle. It is the pivot of the whole system. Hence *Sanyam* on it leads to knowledge of the conditions of the body. We will now come to the other parts of the body. And the first, the pit of the throat. This is the region about the pharynx where the breath from the mouth and the nostrils meets. It is said that the contact of *Pran* with this region produces hunger and thirst, which therefore may be checked by performing *Sanyam* on this part to neutralize the effects of the contact. It may be remarked that the fifth nerve-circle called *Vishudhchakr* is situated somewhere about the same region and anyone who is able to concentrate his breath in that circle and upward easily acquires freedom from hunger and thirst besides other powers. Next we come to the Koormnaree or the nerve where in the breath called *Koorm*, and *Sanyam* on this leads to such a fixate of the body as to make it completely steady and immoveable.

Next there is the light in the head, *i.e.* the collective flow of the light of *Satv* which is seen at the *Brahamranghr* which is variously supposed to be somewhere near the coronal artery, the pineal gland or over the medulla oblongata. Just as the light of a house presents a luminous appearance at the keyhole, so even does the light of *Satv* show itself at the crown of head. This light is very familiar to all acquainted even slightly with *Yoga* practices and is seen even by concentration on the space between the eyebrows. By *Sanyam* on this light is acquired the sight of the yogis called *Siddha*, *i.e.* experts in such wonderful sciences so that (sciences with the aid of which?) you can see things not withstanding the obstacles of space and other things. But the real object of *Yoga* seems to obtain the prefect intuitive power which results from *Pratibha*. *Pratibha* is that degree of intellect which develops itself without any special cause and which is capable of leading to real knowledge. It corresponds to what is generally called intuition. If the yogi tries simply to develop this faculty in him by performing *Sanyam* on the intellect he becomes able to accomplish all that we have referred to before, only through the help of *Pratibha*. This sort of *Pratibha* is called *Tarak gyan* the knowledge that saves *i.e.* leads to final absolution *Moksh*. Hence that *Yoga* which entirely concerns itself with this department of intellectual and spiritual development is often called *Tarakyog* or *Rajyog*.

We come to other parts of the body. *Sanyam* on heart, by which is meant a nerve-circle called *Anahat*, leads to a knowledge to the mind of others as well as one's own.

33. *Sanyam* on the *Purush*-soul itself-leads to the knowledge of soul. The *Samkhya* as well as the *Yoga* lays great stress on the point that *Sanyam*, the source of intelligence, is apart and distinct from the ultimate essence of consciousness. The theory is that *Purush* being reflected in the clear *Satv* enlivens it, and all experience is assumed by the *Satv* so enlivened to be entirely its own act. This confused identification of the two, ever distinct by nature, is the cause if all varied experience. The experience, which the Satv receives, is of no use to itself. It is all for *Purush*; for all the actions of *Prakriti*, which is the source of Satv, and the correlative of *Purush* is for *Purush*. Hence the action of *Satv* is for another and not for itself. Therefore the *Sanyam* on self; *i.e.* on *Purush* right nature and purpose, will lead to a clear knowledge of *Purush*. And thence is produced cognition without the intervention of the organs of sense, *i.e.* intuition cognition of sound, touch, light, taste and smell. The wonderful or, if we may choose to say so, occult powers described hitherto are often all positive obstacles in the way of *Samadhi*, *i.e. Yoga* proper whose nature and import is that state in which the soul sees itself. The author of the *Yoga Sutras* distinctly says that the occult powers serve as obstacles because they become the cause of distracting the mind by the various feelings they excite. Of course, they are not quite useless in as much as they are powers for good in moments when *Samadhi* is suspended. After all, so far as the *Samadhi* highest spiritual aim is concerned, and certainly that is the aim of all philosophies, the exercise of these powers is a positive obstacle on the way to *Samadhi*. This is clearly stated in the *Yoga Sutras*. But in the *Yoga* aphorisms published by Mr. Judge of New York this portion is mistranslated (See Judge's *Yoga Sutras*).

34. The breath in the body is divided into five classes. The air intercepted between the tip of the nose and heart is called *Pran*,that between the heart and the navel is called, *Sman*, that from the navel to the toes of the feet is called *Upan*, that above the tip of nose is called *Udan* and the which pervades the whole body is called *Vyan*. Their respective functions are vitalizing, digestions, expulsion of the excrements, raising up the sound etc. and motion is general. The *Udan* air has the tendency to raise the body upward and carry it above water etc. Hence by mastery over *Udan* there arises the power of ascension, non-contact with water, mud, thorn etc. With reference to the *Sman* breath, the part about the navel is this seat where it performs the function of digestion by keeping the internal fire. When *Sanyam* is performed on *Sman*, this fire can be seen about the whole body which will on that account appear effulgent. This effulgence is most perceptible about the head, between the eye-brows and at the navel. It is said to be the basis of the magnetic currents of living beings.

35. By *Sanyam* on the relation between ether which is the substratum of sound-vibrations and the sense of hearing arises the power of clair-audience. By *Sanyam* on the relation between the body and the *Akash* (ether) arises the power of passing through endless space. And There are many other powers which the *Yoga* claims can arise by reason of performing *Sanyam* on different things.

36. The true yogi does not attach himself to these occult powers. And the *Yoga Sutra* expressly says in one of its aphorisms that it is by non-attachment to this that *Kaivalya* the highest spiritual knowledge is attained.

36. 37. We will now discuss this final aim of *Yoga*. In doing so we will have to refer once more to the nature and doings of *Prakriti*. There are many hundred points for which the *Yoga* philosophy offers its solution. For instance, how is one body changed into another is that the flow of *Prakriti* does it all, the flow of *Prakriti i.e.* that inscrutable action of matter which performs all the work of transformation as seen in the material universe. The very potencies of matter do all and by powerful application produce the necessary conditions for independent

action. The incidental cause in the production of material results, are our virtuous and vicious actions. It may be asked if *Prakriti* does all by its action and produces transformations equal to its potentialities, where is the use of individual good or bad actions. The performance of such acts is not useful in setting up the action of *Prakriti* but it only prepares the way for its free action by removing if good the obstruction in its way. An illustration in point is that of a husband's man who only removes the obstacles in the way of the water which then passes of itself from one spot to another. If the performance of good acts removes all obstacles and prepares the way for the free action of *Prakriti*, a *yogi*, whose vision reveals to him all he has still to go through, may with, as it were, to multiply himself and thus undergo at one and the same time the fruition of all that is to happen. In this he would require, as many minds as there are bodies and the question would arise, whence do these come, it being taken for granted that a yogi can duplicate his gross body. Such a yogi has full command over *Mhat the* root of all egoism and everything else, which makes up "mind". The sense of being or individuality is the result of *Mhat* and the yogi who has command over it is able to send forth as many minds as he likes from this grand reservoir. And as the one mind of the yogi is the cause of all the minds in their various activities the same individual is preserved in all the different bodies with different minds. These newly created minds are not susceptible to impressions, they being produced by means of *Samadhi* and because *yogis* do not acquire impression by actions. Actions or *Karma* are considered under four heads; white, black mixed and indifferent. The first are of gods, the second of wicked beings, the third of men and the fourth of yogis. In other words, yogis acquire no impression by their acts, for they are perfect in non-attachment and hence are ever considered free. From the first three kinds of *Karma* those impressions alone are developed for which the conditions are favorable. In other words, every act leaves an impression and these are collected one upon the other, and new ones added to them as any of them spends itself away by producing its proper result under proper conditions. Only those impressions manifest themselves for which conditions are favourable. For example, if a being who is a man becomes a man again, after passing through the dog, the wolf and the ape, it is certain that such impressions alone will manifest themselves in each or any of these existences as are favored by the conditions.

38. Patajali the author of *Yoga Sutras* discusses from these facts many metaphysical points about the nature of mind and soul. We will however come at once to the final emancipation or *Kaivalya*. And first the qualifications of one, who attains to it. One who has the desire to know what the soul is and what relation his mind and the universe bear to it is said to be desirous of *Kaivalya*. When such a person clearly experiences the distinction between mind and soul and understands the power and nature of either the said desire is distinguished within him. *Kaivalya* is in fact a state in which there is entire cessation of all desire and when the nature of the essence of all consciousness is known there is no room for any action of the mind the source of phenomena. The mind before such knowledge was bent towards worldly objects but now it is entirely bent on discrimination knowledge. This knowledge is of the kind of clear cognition of the different between mind and soul. Not only this but mind is entirely, full of the idea of *Kaivalya* to the exclusion of other thoughts. But while the condition of entire devotion to *Kaivalya* is suspended, there are other thoughts from previous impression or impressions of previous births. These impressions are to be destroyed like other distraction. Even full discrimination is not the desired end and should be suspended by supreme non-attachment which is the nearest road to *Samadhi*, the door of *Kaivalya*. From constant discriminative recognition of the 26 elements of this philosophy results the lights of knowledge; after this the *yogi* works entirely without attachment to any object of desire; then he reaches the state of supreme non-attachment, wherein the lights of soul breaks out in full. In fact all appears full of soul and there is nothing to interrupt this blissful perception. Then all distortion and action cease altogether at least for the yogi. When the distraction are destroyed and when *Karma* and is rendered powerless for good or for ill, there arises full knowledge which is free from the obscuration caused by *Rajas* and *Tames*

and cleared of all impurities arising from the distractions. This knowledge is infinite. As compared to this infinity, that which ordinary men regard as knowable appears but as insignificantly small things. It is easy to know it any time though is not possible that the desire to know a comparatively worthless thing should ever arise.

While such knowledge arises and supreme non-attachment is at highest there arises in the *yogi* entire cessation of the effects of three Gun, the properties. The properties work for the *Purush*; the *Purush* having known himself the properties cease to act, they having fulfilled their end. The whole universe is but a succession of transformation upon transformation of properties. These transformations take an inverse source till all is reduced to matter with the three qualities. No fresh transformations comes take place and hence the succession of transformation comes to an end on the case of the *Purush* who has understood *Kaivalya*. Their effects the various transformation merge onto the higher source and nothing remains for the *Purush* to cognize. This state of the *Purush* is *Kaivalya* or the state of singleness. It dose not mean that the universe is reduced to nothing, for it continues to exist for all those who have not acquired knowledge.

In the case of one who has not acquired knowledge, the visible universe, the cause of distraction, the state of concentration, the supreme idea of non-attachment, all with their impression merge into the mind, which again merges into mere being, which resolves itself in *Mahat*, which finally loses itself in *Prakriti*. This *Kaivalya* of *Prakriti* is by way of metaphor said to be of *Purush*. Or *Kaivalya* may be explained from the side of the *Purush*. When the *Purush* has so far received due illumination as to estrange itself from all relation with *Prakriti* and its transformations it is said to be *Kaivalya (Kaival)* alone or in a state of *Kaivalya*. This is the power of soul centered in itself. *Kaivalya* is not any state of negation or annihilation as some are misled to think. The soul in *Kaivalya* has his sphere of action transferred to a higher plane limited by a limitless horizon. This, our limited minds cannot hope to understand.

The Naya Philosophy

1. Having finished our discussion on the *Samkhya* and its counterpart the *Yoga* philosophy we now enter upon the *Naya* of Gautama with its supplement the *Vaisheshika*.

The author or rather the recognized promulgator of the *Naya* philosophy is Gautama. This philosophy starts with the proposition that in order to obtain the *summum bonum* one must acquire the knowledge of the truth; knowledge of the truth drives away miseries, births, mundane existence, faults and false knowledge and the result is *Moksha*, the freedom of the soul. How can the knowledge of the truth be obtained? Gautama says: 'Knowledge of sixteen topics leads to *Moksha*. What are these sixteen topics? They are all connected with the process of reasoning and the laws of thought. We do not find in *Naya* any prominence given to the rational demonstration of the universe. This we shall find in its complement the *Vaisheshika*. The *Naya* therefore teaches us the method of investigation, the *Vaisheshika* following that method actually tries to investigate into the nature of the universe.

2. The *Naya* mode of investigation may seem very peculiar to those who are not acquainted with the Hindu mode of thinking but it is quite Indian and unique. It says that if you wish to investigate into the nature of things you must proceed first to mention *Udaish*, then give the *Lakshan* of those things and lastly to make *Pareeksha*. I shall explain these terms. First you have to mention *Udaish*, *i.e.* only to name the things by their respective names. Then you have to give the Lakshan of those things, *i.e.* give the differentia of those things-differentia, *i.e.* those qualities which belong to them only and to nothing else and which at the same time are their essential qualities, *i.e.* qualities without which they cannot exist. This means that after naming them you have to give their logical definitions. And thirdly you have to examine whether those definitions are right. The sixteen topics of *Naya* philosophy are treated in that way. We shall proceed with them in order.

3. The first is *Prman*, *i.e.* the means or instruments by which *Pram* or the right measure of any subject is to be obtained.

These are the different processes by which the mind arrives at a true and accurate knowledge. These processes are four *Prtyaksh, Anuman, Upman,* and *Shabd*. We shall describe them when we come to the *Vaisheshika* philosophy. The second topic is *Prmaiya* by which is meant all the objects or subjects of right knowledge. They are twelve in number: *Atma* (soul), Shreer (body), Indriyas (organs or senses), *Arth* (objects of sense), *Buddhi* (understanding or intellect), *Man* (mind), *Prvriti* (activity), *Dosh* (faults), *Praityabhav* (transmigration), *Phal* consequences or fruits), *Dukha* (pain), *Apvarg* (emancipation). These are the twelve of which we have to get the right knowledge by any one of the four processes. The other fourteen topics are not different categories under which things can be classed but rather regular stages through which a logical controversy is to pass. For instance, in discussing a topic there is first the state of *Sanshaya* or doubt about the point to be discussed. Next there must be a *Pryojan* or motive for discussing it. Next a *drishant* or a familiar example must be adduced in order that a *Sidhant* or established conclusion may be arrived at. These four with the former two *Prman* and *Prmaiya,* make up six. The seventh is *Avayava*, *i.e.* the argument of the objector split up. The eighth is *Tark* or refutation of his objection. The ninth is *Nirnaya* or coming to a conclusion. But this is not enough for the *Naya* philosopher. He thinks that every side of a question must be examined, every possible objection stated and so a further *Vad* or controversy takes place which of course leads to *Jalpa* (mere wrangling), followed by *Vitanda (*caviling), *Haitvabhas* (fallacious reasoning), *Chhala* (quibbling artifices), *Jati* futile replies), and *Nigrahsthan* (the putting an end to all discussion by a demonstration of the objector's incapacity for argument). These are *Gautama's* sixteen topics.

4. The most important part of the [philosophy] is the *Vaisheshika* system. The *Naya* of Gautama does not aim at a {demonstration of the] universe. The aim of every [philosophy] ought to be to give an [analytical] demonstration of the [universe, it being] the way for obtaining the summum bonum. The *Naya* only mentions the objects or subjects to be known but it is *Kanada*, the author of the *Vaisheshika*, who tries to analyze

the things and then lays down that final liberation-the summum bonum-follows the right understanding of things. His method is that of generalization. He arranges all the nameable objects, their properties or abstractions even, under seven categories. Let us place ourselves in his position and look at the universe as he does; then only we will be able to understand his philosophy.

5. We [observe] things around us; we see uniformity [and variety] in them. What is that [uniformity and what] is that variety? That [something which] is common to many things, which [is all-pervading] and is without beginning or end [accounts for] uniformity. Notwith..... objects, we see variety in them(?); [notwithstanding] common properties found in all of them, there is something which individualizes them. This is variety. The *Vaisheshika* called uniformity or generality *Samanya* and variety or individuality *Vishaish*. They are the same as genus and species. But this generality and individuality do not exist by themselves. They exist in something. That something which is the tabernacle of qualities or energies is what the *Vaisheshika* calls *Dravya* (substance). He thinks that the qualities and energies or actions are separate entities and therefore ought to be classed under separate categories. The first are what he calls *Gun* (Qualities), the second are *Karma* (actions). We saw before that generality or individuality does not exist without a substance; so there must be some intimate relation between them; in the same manner, we do not see qualities or actions except in substances; (so) there must be an intimate relation between substances and their qualities or actions. This relation is classified by the *Vaisheshika* under a separate category and is named *Samvay* or perpetual intimate relation. Thus all the objects can be classed under six heads *Dravya* (substance), *Gun (Quality), Karma* (actions), *Samanya* (generality), *Vishaish* (individuality) and *Samvaye* (the perpetual relation). There is nothing in the universe outside these six categories. In order however to include negative qualities into the nameable objects-as darkness which is the absence of light, a seventh category called *Abhav* or non-existence or negation of existence is added to the six mentioned before.

6. We will now proceed with these categories one by one.

   i. The first is *Dravya* or substance. Kanada divides them into nine classes-*Prithvi* (earth), *Jal* (water), *taijasa* (light), Vayu (air), *Akash* (ether), *Kal* (time), *Dik* (space), *Atma* (soul), *Manas* (mind). These are the nine substances, each existing as an entity. There is no substance, material or spiritual, outside these nine.

   (ii) The second category is *Gun* or quality. According to this philosophy there are only 24 qualities and no more. These are *Roop* (colour), *Res* (savor or taste), *Gandha* (odour), *Sparsh* (tangibility), *Samkhya* (number), *Pariman* (dimension), *Prithkatv* (individuality), *Sanyoga* (conjunction), *Vibhaga* (disjunction, *Pratv* (priority), *Apratv* (posteriority), (intellect), *Sukha* (pleasure), *Dukha* (pain), *Ichha* (desire, *Dvaish* (aversion), *Pryatn* (volition), *Gurutva* (gravity), *Dravatv* (fluidity), *Snaih* (viscidity), *Sanskar* (self-productiveness), *Dharma* (merit),*Adharma* (demerit), and *Shabd* (sound).

   (iii) The third category action is fivefold: *Utkshaipan* (Elevation or throwing upwards), *Avkshaipan* (Depression or throwing downwards), *Akunchan* (contraction), *Sanprsaran* (dilatation), and *Gaman* (motion in general).

   (iv) The fourth category is *samanya* (generality). It is twofold, higher and lower. All the different objects thought different one from each other are known as substance. Their being substance is the highest generalization.

   But these different objects may be divided into several classes, each class differing from the other. All the objects included in one class have a lower generality and so on.

   (v) The fifth category *Vishaish* (individuality) is of infinite nature. Each atom is separate from the other. And therefore there are infinite individualities.

(vi) The sixth category *Samvay* or intimate relation is that which exists between a substance and it qualities, between atoms and, what is formed out of them, between the whole and its parts, between atoms and what is formed out of them, between the whole and its parts, between substance and its modifications.

(vii) The seventh category is non-existence, which is very easy to understand.

7. We will examine these categories a little closer.

(a) Of the nine substances, earth, water, light and air are considered eternal and non-eternal. The atoms of these substances are eternal but their different manifestations are not eternal. With regard to the creation of the universe the *Vaisheshika* supports the atomic theory and states that the material universe is created out of these four elements. The *Vaisheshika* believe in a personal creator because they think that although the elements were here yet there must be some one to form them into different shapes. For the formation of a pot, although the clay is there, still there is the necessity of a potter. By the will of this divine power motion is imparted to the atoms and evolution follows.

(b) Besides these four elementary substances, there are five other substances-ether, time, space, soul and mind. These are eternal and all of them except mind are all-pervading, *i.e.* they exist everywhere. This means that the soul of every man exists as much in Chicago as in Bombay. The mind however is atomic and is connected with soul. When the soul becomes related with mind knowledge is the result; knowledge is a special characteristic of soul, but it is mind, which receives the sensation of pleasure or pain. The different senses are only the instruments of knowledge. The effects of acts are stored in the mind and they manifest themselves as pleasures and pains in future incarnations. When by the grace of god the soul acquires the right knowledge of things all miseries vanish and the supreme bliss follows.

Mimamsa

1. The next school of thought to which we come the *Mimasma* of Jaimini. *Jaimini's* system cannot really be called a philosophy. It is rather a system of ritualism. It dose not concern itself with investigation into the nature of soul, mind, matter but with a correct interpretation of the ritual of the *Veda* and the solutions of doubts and discrepancies in regard to Vedic texts caused by the discordant explanations of opposite schools, It is therefore a critical commentary on the ritual portion of the *Veda*. We shall therefore at once pass to the *Vedanta* philosophy.

The Vedanta Philosophy

1. The whole *Vedanta* philosophy is based on the *Upanishad* portion of the *Vedas*. The *Chhandogya Upanishad* contains several allegories, which have become the starting point of the philosophy.

There is, for example, a dialogue in the Chhandogya Upanishad between a young student Shwetaketu and his father Uddalaka Aruni, in which the father tries to convince the son, that with all his theological learning, he knows nothing and then tries to lead him on to the highest knowledge, the *Tatvmai* or 'thou art that'. The father said to him, "Shwetaketu, go to school, for there is none belonging to our race, darling, who not having studied, is, as it were, a *Brahman* by birth only.

He began his apprenticeship with a teacher when he was 12 years of age. He returned home when he was 24, having then studied all the Vedas-conceited, considering him well-read and very stern. His father said, to him, "Shwetaketu, as you are so conceited, considering yourself so well read and so stern, my dear, have you asked for that instruction by which we hear what is not audible, by which we perceive what is not perceptible, by which we know what is unknowable." "What is that instruction, Sir?" he asked. The father replied,b My dear, as by one clod of clay all that is made a clay is known, the difference being only a name arising from speech; but the truth being that all is clay; and as, my dear, by one nugget of gold all the is made of gold is known, the difference being only a name arising from speech, but the truth being only a name arising from speech, but truth being that all is gold; and, as my dear, by one pair of nail scissors all that is made of iron is known, the difference being only a name arising from speech, but the truth being that all is iron. Thus, my dear, is that instruction." The son said, "Surely those venerable men (my teachers) did not know that. For if they had known it why should they not have told it me? Do you, Sir therefore, tell me that."

The father said, "In the beginning, my dear, there was that only which, is, one only without a second. Others say, in the beginning there was that only which is not, one only without a second; and from that which is not, that which is was born. But how could it be thus, my dear? How could that which is be born of that which is not? No, my dear, only that which is, was in the beginning, one without a second. It thought, may I be many, may I grow forth. It sent forth fire. That fire thought, may I be many, May I grow forth. It sent forth water. Water thought, may I be many, may I grow forth.

It sent forth. It sent forth fire. That fire thought, may I be many, may I grow forth. It sent forth water. Water thought, may I be many, may I grow forth. It sent forth earth (or food). Therefore whenever it rains anywhere, most food is then produced. From water alone is eatable food produced. As the bees, my son, make honey by collecting the juices of descant trees and reduce the juice into one form, and these juices have no discrimination, so that they might say, I am the juice of this tree or that tree, in the same manner, my son, all these creatures when they have become merged in the true (either in deep sleep or death) know not that they are merged in the true. Whatever these creatures are here, whether lion or a wolf or a boar or a worm or a midge or a gnat or a mosquito, that they become again and again. Now that which is the subtle essence, in it all that exists has its self. It is the true. It is the self, and thou, O Shwetaketu, art it."

"Please, Sir, inform me still more", said the son. "Be it so, my child," the father replied. "These rivers, my son, run the eastern like the Ganges to the East, the western like the Indus to the West. They go from sea to sea, *i.e.*, the clouds lift up the water from the sea to the sky and send it back as rain to sea. They become indeed seas. And as those rivers, when they are in the sea, do not know, I am this or that river, in the same manner, my son, all these creatures when they have come back from the true know not that they have come back from the true. Whatever these creatures are here, whether a lion or a mosquito, that they become again and again. That which is that subtle essence, in it all that exists has its self. It is the true. It is the self, and thou, O Shwetaketu, art it."

"Please, Sir, inform me still more," said the son. "Be it so, my child," the father replied. "If some one were to strike at the root of this large tree here, it would bleed but live. If he were to strike at its stem, it would bleed but live. If he were to strike at its tip, it would bleed but live. Pervaded by the living self that tree stand s firm, drinking in its nourishment and rejoicing. But if life (the living self) leaves one of its branches, that branch withers, if it leaves the whole tree, the whole tree withers. In exactly the same manner, my son, know this. This body indeed withers and dies when the living self has left it; the living self never dies. That which is that subtle essence, in it all that exists has its self. It is the true. It is the self, and thou, O Shwetaketu, art it."

"Please, Sir inform me still more," The son said. "Be it so, my child," the father said. Place this salt in water and then wait on me in the morning." the son did as was commanded. The father said to him, "Bring me the salt which you placed in the water last night," The son having looked for it found it not, for of course it was melted. The son having looked for it found it not, for of course it was melted. The father said, "Taste it from the surface of the water. How is it?" The son replied, "It is salt." "Taste it from the middle, How is it?" "It is salt" "Taste it from the bottom. How is it?" The son said, "It is salt." The father said, "Now leave the vessel and sit by my side." He did so. The father asked, "Where is the salt? Do you see it?" The son said, "I do not see it but it is in the water." The father said, "Here also in this body you do not perceive the true, my son, but there indeed it is. That which is the subtle essence, in it all that exists has its self. It is the true. It is the self, and thou, O Shwetaketu, art it."

"Please, Sir, inform me more." "Be it so, my child. If a man is ill, his relatives assemble round him and ask-dost thou know me? Now as long as his speech is not merged in the mind, his mind in breath, his breath in heat, heat in the highest Godhead, he knows them. But when his speech is merged in his mind, his mind in breath, breath in heat, and heat in the highest Godhead, he knows them not. That which is the subtle essence, in it all that exists has its self. It is the true. It is the self, and thou, O Shwetaketu, art it."

2. I told you last time that the *Vedanta* philosophy is based on the *Upanishads*. The ritual of the *Vedas* was considered the *Karmkand* or the work portion, the Upanishads constituted what is known as *Gyankand* or the knowledge portion in so far as it propounds a certain theory of the world. I also told you that Mimamsa was a system of ritualism which gave a correct interpretation of the ritual of the *Veda* and the solutions of doubts and discrepancies in regard to *Vedic* texts caused by the discordant explanations of opposite schools. Just as the ritualistic portion of the *Vedas* became object of comment by *Jaimini*, the author of the *Mimamsa* so did Badarayana comment on or rather composed aphorisms based on the Upanishads.

3. The *Vedanta* philosophy has its two chief supporters Shankara and Ramanuja. Both of them rest their doctrines on the Upanishads. In Shankara's opinion the Upanishads teach as follows:

(a) Whatever is, is in reality one; there truly exists only one universal being called Brahma or *Parmatman* (the highest self). This being is of an absolutely homogeneous nature; it is pure being or, which comes to the same, pure intelligence or thought *Chaetanya, gyan.* Intelligence or thought is not to be predicated of Brahma as its attribute but constitutes its substance. Brahma is not thinking being, but thought itself. It is absolutely destitute of qualities; whatever qualities or attributes are conceivable can only be denied of it. But if nothing exists but one absolutely simple being, whence the appearance of the world by which we see ourselves surrounded and in which we ourselves exist as individual beings?

The answer is that Brahma is associated with a certain power called *Maya* or *Avidya.* This power cannot be called 'being', for 'being' is only Brahma; nor can it be called' not-being' in the strict sense, for it at any rate produces the appearance of this world. It is in fact a principle of illusion, the non-definable cause owing to which there seems to exist a material world comprehending distinct individual existences. Being associated with this principle of illusion Brahma is enabled to project the appearance of the world, in the same way as a magician is enabled by his incomprehensible magical power to produce illusory appearances of animate and inanimate beings. *Maya* thus constitutes the *Upadan* (the material cause) of the world, or if we wish to call attention to the circumstance that *Maya* belongs to Brahma as *Shakti*, we may say that the material cause of the world is Brahma in so far as it is associated with *Maya*. In this latter quality Brahma is more properly called *Ishwar* (the Lord).

Maya under the guidance of the Lord modifies itself by a progressive evolution into all the individual existences distinguished by special names and forms, of which the world consists; from it there spring in due succession the different material elements and the whole bodily apparatus belonging to sentient beings. In all those apparently individual forms of existence the one indivisible Brahma is present, but owing to the particular adjuncts into which ( ) has specialized itself it appears to be broken up-it is broken up, as it were-into a multiplicity of intellectual or sentient principles, the so called *Jeev* (individual or personal souls). What is real in each is only the universal Brahma itself, the whole aggregate of individualizing bodily organs and mental functions, which in our ordinary experience separate and distinguish one *Jeev* from another, is the offspring of *Maya* and as such unreal.

The phenomenal world or world of ordinary experience *Vyvehar* thus consists of a number of individual souls engaged in specific cognition's, volition's and so on and of the external material objects with which those cognition's and volition's are concerned. Neither the specific cognition's nor their objects are real in the true sense of the world, for both are altogether due to *Maya*. But at the same time we have to reject the idealistic doctrine of certain Buddhist schools according to which nothing whatever truly exists but certain trains of cognition acts or ideas to which no external objects correspond for external things, although not real in the strict sense of the word, enjoy at any rate as much reality as the specific acts, whose objects they are.

(b) The non-enlightened soul is unable to look through and beyond *Maya*, which like a veil hides from it its true nature. Instead of recognizing itself to be Brahman it blindly identifies itself with its adjuncts *Upadhi*-the fictitious off springs of *Maya*, and thus looks for its true self in the body, the sense-organs and the internal organ *Manas*, *i.e.*, the organ of specific cognition. The soul, which in reality is pure intelligence, non-active, infinite, thus becomes limited in extent, as it were, limited in knowledge and power, an agent and enjoyer. Through its actions it burdens itself with merit and demerit, the consequences of which it has to bear or enjoy in series of future embodied existences, the Lord-as retributer and dispenser-allotting to each soul that form of embodiment to which it is entitled by its previous actions. At the end of each of the great world periods called *Kelp*, the Lord retracts the whole world, *i.e.* the whole material world is dissolved and merged into non-distinct *Maya* while the individual souls, free for the time from actual connection with *Upadhi*, lie in deep slumber as it were. But as the consequences of their former deeds are not yet exhausted they have again to enter an embodied existence as soon as the Lord sends forth a new material world, and the old round of birth, action, death begins anew to last to all eternity as it has lasted from all eternity.

(c) The means of escaping from this endless *Sansara*, the way-out of which can never be found by the non-enlightened soul, are furnished by the *Veda*. The *Karmkand* indeed whose purport it is to enjoin certain actions cannot lead to final release, for even the most meritorious works necessarily lead to new forms of embodied existence. And in the *Gyankand* of the *Veda* also two different parts have to be distinguished, *viz.* firstly those chapters and passages which treat of Brahma in so far as it is related to the world and hence characterized by various attributes, *i.e.* of *Ishwar* or lower Brahma, and secondly, those texts which set forth the nature of the highest Brahma transcending all qualities and the fundamental identity of the individual soul with that highest Brahma. Devout meditation on Brahma as suggested by passages of the former kind does not directly lead to final emancipation; the pious worshipper passes on his death into the world of the lower Brahma only, where he continues to exist as a distinct individual soul although in the enjoyment of great power and knowledge-until at last he reaches the highest knowledge and through it final release. That student of the *Veda*, on the other hand, whose soul has been enlightened by the texts embodying the higher knowledge of Brahma, whom passages such as the great saying 'That art thou' have taught that there is no difference between his true self and the highest self, obtains at the moment of the death immediate final release, *i.e.* he withdraws altogether from the influence of *Maya* and asserts himself in his true nature which is nothing else but the absolute highest Brahma. This is the teaching of Shankara.

4. According to Ramanuja, on the other hand, the teaching of the Upanishads is a little different.

a. He says: There exists only one all-embracing being called Brahma or the highest self or the Lord. This being is not destitute of attributes but rather endowed with all imaginable auspicious qualities. It is not intelligence as Shankara maintains but intelligence is its chief attribute. The Lord is all-pervading, all-powerful, all knowing, all merciful; his nature is fundamentally antagonistic to all evil. He contains within himself whatever exists. While according to Shankara, the only reality is to be found in the nonqualified homogenous highest Brahma which can only be defined as pure being or pure thought, all plurality being a mere illusion, Brahma according to Ramanuja's view comprises within itself distinct elements of plurality which all of them lay claim to absolute reality of one and the same kind. Whatever is presented to us by ordinary experience, *viz.* matter in all its various modifications and the individual souls of different classes and degrees, are essential, real constituents of Brahma's nature. Matter and souls *Achit and Chit* constitute according to Ramanuja's terminology the body of the Lord, they stand to him in the same relation of entire dependence and subservience in which the matter forming an animal or vegetable body stands to its soul or animating principle. The Lord pervades and rules all things which exist-material or immaterial-as their *Anteryamee* the fundamental text for this special Ramanuja's tenet-which in the writing of the sect is quoted again and again-is the so called *Anteryamee Brahman*, which says that within all elements, all sense organs and lastly within all individual souls there abides an inward ruler whose body these elements, sense organs an individual souls constitute. Matter and souls as forming the body of the Lord are also called modes *Prkar* of him. They are to be looked upon as his effects, but they have enjoyed the kind of individual existence, which is theirs from all eternity and will never be entirely resolved in Brahma. They however exist in two different periodically alternating conditions. At some time they exist in a subtle state in which they do not possess those qualities by which they are ordinarily known, and there is then no distinction of individual name and form. Matter in that state is non-evolved *Avyakt* individual souls are not joined to material bodies and their intelligence is in a state of contraction *Sankoch*.

This is the *Prley* State, which recurs at the end of each *Kalpa*, and Brahma is then said to be in its causal condition *Karn·avastha*. To that state all those *Vedic* passage refer which speak of the Brahma or self as being in the beginning one only without a second. Brahma then is indeed not absolutely one, for it contains within itself matter and soul in a germinal condition; but as in that condition they are so subtle as not to allow of individual distinctions being made, they are not counted as something second in addition to Brahma. When the *Prley* state comes to an end, creation takes place owing to an act of volition on the Lord's part The primary non-evolved matter then passes over into its other condition; it becomes gross and thus acquires all those sensible attributes, visibility, tangibility and so on, which are known from ordinary experience. At the same time the souls enter into connection with material bodies corresponding to the degree of merit or demerit acquired by them in previous forms of existence; their intelligence at the same time undergoes certain expansion *Vikas*. The Lord together with matter in its gross state and the expanded souls is Brahma in the condition of effect *Karyavstha*. Cause and effect are thus at the bottom the same; for he effect is nothing but the cause, which has undergone a certain change *Parin*·am. Hence the cause being known, the effect is known likewise.

(b) Owing to the effects of their former actions the individual souls are implicated in the *Sansara*, the endless cycle of birth, action and death, final escape from which is to be obtained only through the study of the *Gyankand* of *Veda*. Compliance with the *Karmkand* does not lead outside the *Sansara*. But he who, assisted by the grace of the Lord, cognizes and meditates on him in the way prescribed by the *Upanishads* reaches at his death final emancipation, *i.e.* he passes through the different stages of the path of the Gods up to the world of Brahma and there enjoys an everlasting blissful existence from which there is no return into the sphere of transmigration. The characteristics of the released soul are similar to those of Brahma; it participates in all the latter's glorious qualities and powers, excepting only Brahma's power to emit, rule and retract the entire world.

5. The chief points in which the two systems agree on the one hand and diverge on the other are these: Both systems teach *Advaet i.e.* non-duality or monism. There exist not several fundamentally distinct principles, such as *Prakriti* and *Purush* of the *Samkhya*, but there exists only one all-embracing being. While, however, the *Advaet* taught by Shankara is a rigorous, absolute one, Ramanuja's doctrine has to be characterized as

Vishishtadvaesh i.e. qualified non-duality, non-duality with a difference. According to Shankara, whatever is, is Brahma, and Brahma itself is absolutely homogeneous, so that all difference and plurality must be illusory. According to Ramanuja also, whatever is, is Brahma, but Brahma is not of homogeneous nature, but contains within itself elements of plurality, owing to which it truly manifests itself in a diversified world with its variety of material forms of existence and individual souls is not unreal *Maya* but a real part of Brahma's nature, the body investing the universal self. The Brahma of Shankara is in itself impersonal, a homogeneous mass of objectless thought, transcending all attributes; a personal God it becomes only through its association with the unreal principle of *Maya*, so that, strictly speaking, Shankara's personal God, his *Ishwar*, is himself something unreal. Ramanuja's Brahma, on the other hand, is essentially a personal God, the all-powerful and all wise ruler of a real world permeated and animated by his spirit. There is thus no room for the distinction between a *Pram Nirguna* And *Apram Saguna* Brahma, between Brahma and Ishwar. Shankara's individual soul is Brahma in so far as [it is] limited by the unreal *Upadhi* due to *Maya*. The individual soul of Ramanuja, on the other hand, is really individual soul of Ramanuja, on the other hand, is really individual; it has indeed sprung from Brahma and is never outside Brahma, but nevertheless it enjoys a separate personal existence and will remain a personality for ever. The release from *Sansara* means according to Shankara the absolute merging of the individual soul in Brahma, due to the dismissal of the errouneous notion the soul is distinct from Brahma; according to Ramanuja it only means the soul's passing from the troubles of earthly life into a kind of paradise where it will remain for ever in undisturbed personal bliss. As Ramanuja does not distinguish a higher and lower Brahma the distinction of a higher and lower knowledge is likewise not valid for him; the teaching of the Upanishads is not two fold but essentially one, and leads the enlightened devotee to one result only.

6. As Shankara's views are mostly considered to be true, we will follow him in some details as to what he says in his

comments on the *Vedanta* aphorisms. The whole work is divided into four parts, each part containing four parts, each part containing four chapters. We will deal with them in order.

(a) In the first chapter he deals with certain passages from Upanishads referring to the word Brahma. We will consider only that part wherein the word Brahma is defined. Brahma is that from which the origin, subsistence and dissolution of this world proceed. Shankara explains this definition by saying that omniscient omnipotent cause from which proceed the origin, subsistence and dissolution of this world-which world is differentiated by names and forms, contains many agents and enjoyers, it the abode of the fruits of actions, these fruits having their definite places, times and causes, and the nature of whose arrangement cannot even be conceived by mind-that cause is Brahma.

(b) *Vedanta* philosophy then rests on the fundamental conviction of the *Vedantistss* that the Soul and absolute Being or Brahma is one in their essence. In the old Upanishads this conviction rises slowly; but when once it was recognized that the Soul and Brahma were in their deepest essence one, the old mythological language of the Upanishads was given up; for instance the passage representing the soul as travelling on the road of the fathers *Pitryan* or the road of the Gods *Devyan*. We read in the *Vedanta* aphorisms that this approach to the throne of Brahma has its proper meaning so long only as Brahma is still considered personal and endowed with various qualities but that when the knowledge of the true, the absolute and unqualified Brahman, the Absolute Being, has once risen in the mind these mythological concepts have to vanish. "How would it be possible," Shankara says, "that he who is free from all attachment, unchangeable and unmoved, should approach another person, should move or go to another place? The highest oneness, if once truly conceived, excludes anything like an approach to a different object or to a distant place."

(c) The Sanskrit language has the great advantage that it can express the difference between the qualified and the unqualified Brahma by a mere change of gender; Brahma being used as a masculine when it is meant for the qualified and

Brahma as a neuter when it is meant for the unqualified Brahma, the Absolute Being. This is a great help and there is nothing corresponding to it in English.

(d) We must remember also that the fundamental principle of the *Vedanta* philosophy was not "Thou art He' but 'Thou art that' and that it was not 'thou will be' but 'thou art'. This 'thou art' expresses something, that is, that has been and always will be, not something that has still to be achieved, or is to follow, for instance, after death.

Thus Shankara says: "If it is said that the Soul will go to Brahma, that means that it will in future attain, or rather, that it will be in future what, though unconsciously, it always has been, *viz.* Brahma. For when we speak of some one going to some one else, it cannot be one and the same who is distinguished as the subject and the object. Also, if we speak of worship, that can only be if the worshipper is different from the worshipped. By true knowledge the individual soul does not become Brahma but is Brahma as soon as it knows what it really is and always has been. Being and knowing are one here."

(c) Here lies the characteristic difference between *Yoga* philosophy and *Vedanta*. In *Yoga* the human soul is represented as burning with love for God, as filled with a desire for union with or absorption in God. We find little of that in the Upanishads, and when such ideas occur they are argued away by the *Vedanta* philosophers. They always cling to the conviction that the Divine has never been really absent from the human soul, that it always is though covered by darkness or nescience, and that as soon as that darkness or that nescience is removed the soul is once more and in its own right what it always has been. It is-it does not become-Brahma.

(f) Last time I gave you the dialogue from the Chhandogya Upanishad between a young student Shwetaketu and his father. In that dialogue we have only a popular and not yet systematized view of the *Vedanta*. There are several passages indeed, which seem to speak of the union and absorption of the soul rather than of its recovery of its true nature. Such passages are always explained away by the stricter Vedanta philosophers and they have no great difficulty in doing this. For there remains always the explanation that the qualified personal Brahma in the masculine gender is meant and not yet the highest Brahma, which is, free from all qualities. That modified personal Brahma exists for all practical purposes, till its unreality has been discovered through the discovery of the highest Brahma; and as in one sense the modified masculine Brahma is the highest Brahma as soon as we know it and shares all its true reality with the highest Brahma as soon as we know it, many things may in a less strict sense be predicated of Him, the modified Brahma, which in truth apply to it only, the highest Brahma. This amphibole runs through the whole of the *Vedanta* Sutras and a considerable portion of the *sutras* is taken up with the task of showing that when the qualified Brahma seems to be meant it is really the unqualified Brahma that ought to be understood. Again, there are ever so many passages in the Upanishads which seem to refer to the individual soul but which, if properly explained, must be considered as referring to the highest *Atman* that gives support and reality to the individual soul. This at least is the view taken by Shankara, whereas the fact is that there have been different stages in the development of the belief in the highest Brahma and in the highest *Atman*; and some passages in the Upanishads belong to earlier phases of Indian thought when Brahma was still conceived simply as the highest deity and true blessedness was supposed to consist in the gradual approach of the soul to the throne of God.

(g) The fundamental principle of Vedanta philosophy that in reality there exists and there can exist nothing but Brahma, that Brahma is everything, the material as well as the efficient cause of the universe, is of course in contradiction with our ordinary experience. In Indian as any where else, man imagines at first that he in his individual bodily and spiritual character is something that all objects of the outer world also exist as objects. Idealistic philosophy swept this distinction with the *Vedantistss*.

(h) The *Vedanta* philosopher however is not only confronted with this difficulty but he has to meet another difficulty peculiar to himself. The whole of the *Veda* is in his eyes infallible, yet that *Veda* enjoins the worship of many Gods and even in enjoining the worship *Upasana* of *Brahma*, the highest deity in his active masculine and personal character, it recognizes an objective deity different from the subject that is to offer worship and sacrifice to him.

Hence the *Vedanta* philosopher has to tolerate many things. He tolerates the worship of an objective Brahma as a preparation for the knowledge of the subjective and objective or the Absolute Brahma, which is the highest object of his philosophy. He admits one Brahma endowed with quality, but high above the usual Gods of the *Veda*. This Brahma is reached by the pious on the path of the Gods; he can be worshipped and it is he who rewards the pious for their good works. Still, even he is in that character the result of *Avidya* (ignorance, nescience), of the same ignorance which prevents the soul of man, the *Atman*, from distinguishing itself from its encumbrances, the so-called *Upadhis* such as body, the organs of sense and their works.

(i) This nescience can be removed by knowledge only and this knowledge is imparted by the *Vedanta* which shows that all our ordinary knowledge is simply the result of ignorance or nescience, is uncertain, deceitful and perishable or, as we should say, phenomenal, relative and conditioned. The true knowledge called *Smyagdarshan* or complete insight cannot be gained by sensuous perception *Prtyaksh* or by inference *Anuman*, nor can obedience to the law of the *Veda* produce more than temporary enlightenment or happiness. According to the orthodox Vedanta, *Shruti* alone or what is called revelation can impart that knowledge and remove that nescience which is innate in human nature.

(j) Of the higher Brahma nothing can be predicated but that it is and that through our nescience it appears to be this or that.

When a great *Vedantistss* was asked to describe Brahma, he was simply silent-that was his answer. But when it is said that Brahma is, that means at the same time that Brahma is not, that is to say, that Brahma is nothing of what is supposed to exist in our sensuous perceptions.

There are two other qualities, which may safely be assigned to Brahma, namely, that it is intelligent and that it is blissful, or rather that it is intelligence and bliss. Intelligent seems the nearest approach to Sanskrit *Chit* and *Chaetanya*. Spiritual would not answer, because it would not express more than that it is not material. But *Chit* means that it is, that it perceives and knows, though as it can perceive itself only we may say that it is lighted up by its own light or knowledge, or, as it is sometimes expressed, that it is pure knowledge and pure light. We can best understand it when we consider what is negatived by it, namely, dullness, deafness, darkness and all that is material. In several passages a third quality is hinted at, namely blissfulness, but this again only seems another name for perfection and chiefly intended to exclude the idea of any possible suffering in Brahma.

It is in the nature of this Brahma to be always subjective and hence it is said that it cannot be known in the same way as all other objects are known, but only as a knower knows that he knows and he is.

(k) Still whatever is and whatever is known-two things which in the *Vedanta* and in all other idealistic systems of philosophy are identical-all is in the end Brahma. Though we do not know it, it is Brahma that is known to us when conceived as the author or creator of the world, an office, according to Hindu idea, quite unworthy of the Godhead in its true character. It is the same Brahma that is known to us in our own self-consciousness. Whatever we may seem to be or imagine ourselves to be for a time, we are in truth the eternal Brahma, the eternal self. With this conviction in the background, the *Vedantistss* retains his belief in what he calls the Lord, God, the creator and ruler of the world, but only as phenomenal or as adapted to the human understanding. He thinks that just as a man believes in his personal self so he is sure to believe in a personal God, and such personal God may even be worshipped. But we must remember that what is worshipped is only a person, or as the Brahmins call it a *Prteek*, an aspect of the true eternal essence as conceived by us in our inevitably human and limited knowledge. Thus the strictest observance of religion is insisted on while we are what we are. We are told that there is truth in the ordinary belief in God as the creator or cause of the world, but a relative truth only, relative to the human understanding, just as there is truth in the perception of our senses and in the belief in our personality, but relative truth only. His belief in the *Veda* would suffice to prevent the *Vedantistss* from a denial of the Gods or from what we call atheism.

In deference to the *Veda* the *Vedantistss* has even to admit, if not exactly a creation, at least a repeated emanation of the world from Brahma and re-absorption of it into Brahma from *Kelp* to *Kelp* or from age to age.

If we ask what led to a belief in the individual souls the answer we get is the Upadhi, the surroundings or the encumbrances, *i.e.* the body with the breath or life in it, the organs of sense and the mind. These together form the subtle body *Sooksham Shreer* and this *Sooksham Shreer* is supposed to survive while death can destroy the coarse body *Sthool Shreer* only. The individual soul is held by this subtle body and its fates are determined by acts which are continuing in their consequences and which persist in their effects for ever, or at least until true knowledge has arisen and put an end even to the subtle body and to all phantasms of nescience.

(1) How the emanation of the world from Brahma is conceived in *Vedanta* philosophy is of small interest. It is almost purely mythological and indicates a very low knowledge of physical science. Brahma is not indeed represented any longer as a maker or a creator, as an architect or a potter. What we translate by creation *Srishti* means really no more than a letting out and corresponds closely with the theory of emanation. The Upanishads propose ever so many similes by which they wish to render the concept of creation or emanation more intelligible. One of the oldest similes applied to the production of the world from Brahma is that of the spider drawing forth, *i.e.* producing, the web of the world from itself. Another simile, which is meant to do away with what there is left of efficient-besides material-causality in the simile of the spider which after all sill the throwing out and drawing back of the threads of the world, is

that of hair growing from the skull. Nor is the theory of what we call evolution wanting in the Upanishads. One of the most frequent similes used for this is the change of milk into curds. The curds are nothing but the milk only under a different form. It was soon found however that this simile violated the postulate that the One Being must not only be one but that, if perfect in itself, it must be unchangeable. Shankara therefore offered a new theory. It is distinguished by the name of *Vivart* from the *Parin·am* or evolution theory, which is held by Ramanuja. *Vivart* of Shankara means turning away. It teaches that the Supreme Being remains always unchanged and that our believing that anything else can exist beside it arises from *Avidya i.e.* nescience. Most likely this *Avidya* or ignorance was first conceived as purely subjective, for it is illustrated by the ignorance of a man who mistakes a rope for a snake. In this case the rope remains all the time what it is, it is only our ignorance which frightens us and determines our actions. In the same way Brahma always remains the same, it is our ignorance only, which makes us see a phenomenal world and a phenomenal God. Another favorite simile is our mistaking mother-of-pearl for silver. The *Vedantist* says: We may take it for silver but it always remains mother-of-pearl. So we may speak of the snake and the rope, or of the silver and mother-of-pearl, as being one. And yet we do not mean that the rope has actually undergone a change or has turned into silver. After that the *Vedantists* argue that what the rope is to the snake the Supreme Being is to the world, They go on to explain that when they hold that the world is Brahma they do not mean that Brahma is actually transformed into the world, for Brahma cannot change and cannot be transformed. They mean that Brahma presents itself as the world or appears to be the world. The world's reality is not its own but Brahma's, yet Brahma is not the material cause of the world, as the spider is of the web, or the milk of the curds, or the sea of the foam, or the clay of the jar (which is made by the potter), but only the substratum, the illusory material cause. There would be no snake without the rope, there would be no world without the Brahma, and yet the rope does not become a snake nor does Brahma become the world. With the *Vedantists* the phenomenal and the nominal are essentially the same. The silver as we perceive and call it is the same as the mother-of-pearl; without the mother-of-pearl there would be no silver for us. We impart to mother-of-pearl the name and form of silver, and by the same process by which we create silver the whole world was created by worlds and forms.

(m) Besides, the *Vedanta* philosophy has its own theory as to the creation of the whole world out of Brahma and *Avidya*. The purport of the philosophy however comes to this: All being is Brahma, nothing can be except Brahma, while all that exists is an illusory, not a real, modification of Brahma and is caused by name and form. When the true knowledge arises, everything becomes known as Brahma only. We may ask, whence the names and forms and whence the phantasmagoria of unreality. The *Vedantists* has but one answer, it is simply due to *Avidya*. There is another simile. Indian jugglers knew how to make people believe that they saw two or three jugglers while there was only one. The juggler himself remained one, knew himself to be one only-like Brahma; but to the spectators he appeared as many. But all these are similes only and with us there would remain the question whence this nescience. The *Vedantists* is satisfied with the conviction that for a time we are as a matter of fact nescient and what he cares for chiefly is to find out, not how that nescience arose but how it can be removed.

(n) What is the mode of removing this ignorance? *Bharati Tirtha*, a famous *Vedantist*, says: "Neglecting the unreal creation consisting of mere name and form, one should meditate on the Brahma and should ever practice internal as well as external concentration. Internal concentration is of two kinds, *Sviklp*, and *Nirviklp*. The first is the meditation of (on?) the subjective *Atma* as the witness of the mental world-passions, desires etc. arising in the mind. The second is the fixing one's mind on the thought `I am Brahma', [Brahma] which is described in the *Vedas* as self-existent, eternal, all-consciousness and pleasure, self-illumined and unique in itself. That is *Nirviklp* in which, through the ecstasy of the pleasure consequent upon the knowledge of one's self, the sight as well as the world are both overlooked and the mind stands like the jet of a lamp burning in place protected from the slightest breeze. The separation in any external object of sight, of name and form from its original substratum *Sat* is *Drishanuviddh* external concentration. The meditation on the one, unique and *Sachidanand* Brahma as the only reality in the universe is *Shabdanuvridh* external concentration. The third *nirvikalpa* is, like the one described before, cessation of all thought, from the enjoyment of one eternal pleasure. One should devote one's time to these six kinds of *Smadhi*. The false identity of the material shell and the Universal Life being dissolved and the universal *Atman* being thoroughly realized, wherever the mind of the ascetic is directed there it naturally loses itself into one or other of these *Samadhi*. That limit of limits being seen, the knot of *Ahamkar* (egoism) is cut asunder, all doubts disappear, all actions cease to affect."

Buddhism

1. We have described, very shortly though, those schools of philosophy who take *Vedas* as their guide. We are now entering upon another school-one of the two, which have discarded the *Veda*s and followed their own lines of thought. Buddhism is one of them. A philosophy is not born in a day and therefore to say that Buddha while sitting under the Bo tree was inspired as it were with the truths which he afterwards circulated has no meaning. Truths are not reached in a moment. Sciences and arts are not discovered in a day and therefore Buddha who was a Hindu by birth and a follower of the Brahma faith must have been the outcome of his time.

Six centuries before Christ, India witnessed the commencement of a great revolution. The Brahmanical religion had been practiced and proclaimed for centuries of years. The Gods of the Rig-*Veda* whom the ancient had invoked and worshipped lovingly and fervently had come to be regarded as so many names and *Indra* and *Usha* raised no distinct ideas and no grateful emotions. The simple libations of the *Som* juice, which the old *Rishi* had offered to their gods, had developed into cumbrous ceremonials, elaborate rites and utter sacred prayers for the people. The people were taught to believe that they earned merit by having these rites performed and prayers uttered by hired priests.

It was Buddha who created a reaction in such society.

2. About 100 miles northeast of the city of Benares was situated about 600 years before Christ a place called Kpilavastu on the bank of the River Rohini. And two kindred clans-the Shakyas and Kolians-lived on the opposite banks of that river. Kapilavastu was the capital of the Shakyas who were then living in peace with the Kolians and Shuddhodana the king of Shakyas had married two daughters of the king of Kolians. Neither queen bore any child of Shuddodana for many years, and the hope of leaving an heir to the principality of the Shakyas was well nigh abandoned. At last however the elder queen promised her husband an heir and according to ancient custom left for he father's house in order to be confined. But before she reached the place she gave birth to a son in the pleasant grove of *Lumbini*. The mother and child were carried back to Kapilavastu where the mother died 7 days after leaving the child to be nursed by his stepmother and maternal aunt, the younger queen.

The birth of Gautama is naturally the subject of many legends, which have most remarkable resemblance with the legends about the birth of Jesus Christ. The boy was named Siddhartha but Gautama was his family name. He belonged to the Shakya tribe and is therefore called *Shakyasingh*; and when he had proclaimed and preached a reformed religion he was called Buddha or the awakened or enlightened.

Little is known of the early life of young Gautama except that he was married to his cousin Yashodhara, daughter of the king of Koli about the age of 18. It is said that Gautama neglected the manly exercises which all Kshatryas of his age delighted in, and that his relations complained of that. A day was accordingly fixed for the trial of his skill and the young prince of the Shakyas proved his superiority to his kinsmen.

Ten years after his marriage Gautama resolved to quit his home and his wife for the study of philosophy and religion. The story which is told of the young prince abandoning his home and his position is well known. He must have for a long time pondered deeply and sorrowfully on the sins and sufferings of humanity, he must have been struck with the vanity of wealth and position. It is said that the sight of decrepit old man, of a sick man, of decaying corpse and of dignified hermit led him to form his resolution to quit home.

At this time a son was born to him. It is said that the news was announced to him in a garden on the riverside and the pensive young man only exclaimed: This is a new and strong tie I shall have to break." That night he repaired to the threshold of his wife's chamber-and there by the light of the flickering lamp, he gazed on a scene of perfect bliss. His young wife lay surrounded by flowers and with one hand on the infant's head. A yearning arose in his heart to take the babe in his arms for the last time before relinquishing all earthly bliss. But this he might not do. The mother might be awakened and the importunities of the fond and loving soul might unnerve his heart and shake his resolution. Silently then he tore himself away from that place. In that one eventful moment, in the silent darkness of that night he renounced for ever his princely fame and more than all this the affection of happy home, the love of a young wife and of a tender infant now lying unconscious in sleep. He renounced all this and rode away that night to become a poor student and homeless wanderer. His faithful servant Channa asked to be allowed to stay with him and become an ascetic but Gautama sent him back and repaired alone to Rajhagriha.

Rajgrha was the capital of Bimbisara, the king of Magdhas and was situated in a valley surrounded by five hills. Some Brahmin ascetics lived in the caves of this hill sufficiently far from the town for studies and contemplation and yet sufficiently near to obtain supplies. Gautama attached himself first to one Alara and then to another Udraka and learned from them all that Hindu philosophy had to teach.

Not satisfied with this learning Gautama wished to see if penances could bring superhuman insight and power as they were reputed to do. He retired therefore to the jungle of Uruvela near the site of the present temple of Buddha Gaya and for six year attended by five disciples he gave himself up to the severest penance and self-mortification, but he could not obtain what he sought. At last one day he fell down from sheer weakness and his disciples thought he was dead. But he recovered and despairing of deriving any profit from penance he abandoned it. His disciples who did not understand his object lost all respect for him when he gave up his penances. They left him alone and went away to Banaras.

Left alone in the world, Gautama wandered towards the banks of Niranjara, received his morning meal from the hands of Sujata, a village daughter, and set himself down under the Bo-tree or the tree of wisdom. For a long time he sat in contemplation and scenes of his past life came thronging into his mind. The learning he had acquired had produced no results the penances he had undergone were vain, his disciples had left him alone in the world. Would he now return to his loving widowed wife, to his little child now a sweet boy of six years, to his affectionate father and his loyal people? This was possible, but where would be the satisfaction? What would become of the mission to which he had devoted himself? Long he sat in contemplation and doubt, until the doubts cleared away like mists in the morning and the daylight. Truth flashed before his eyes. What was this truth which learning did not touch and penances did not impart? He made no new discovery, he had acquired no new knowledge, but his pious nature and his benevolent heart told him that a holy calm life and love towards others were panacea for all evils. Self-culture and universal love-this was his discovery, this is the essence of Buddhism.

The conflict in Gautam's mind, which thus subsided in calm, is described in Buddhist writings by marvelous incidents. Clouds and darkness prevailed the earth and oceans quaked, rivers flowed back to their sources and peaks of lofty mountains rolled down.

Gautam's old teacher Alara was dead and he therefore went to Benaras to proclaim the truth to his five former disciples. On the way he met a man of the name of Upaka belonging to the Ajivaka sect of ascetics who, looking at the composed and happy expression on Gautam's face asked: "Your countenance, friend, is serene, your complexion is pure and bright. In whose name, friend, have you retired from the world? Who is your teacher and what doctrine do you profess?" To this Gautama replied that he had no teacher, that he had obtained *Nirvana* by the extinction of all passions and added: I go to the city of Kashi to beat the drum of the immortal in the darkness of the world." Upaka did not understand him and replied after a little conversation, "It may be so, friend," shook his head and took another road and went away.

At Benaras, Gautama entered the Deer Park *Migdav* in the cool of the evening and met his former disciples. And he explained to them his new tenets:

"There are two extremes, O Bhikkhus, which the man who has given up the world ought not to follow the habitual practice, on the one hand, of those things whoše attraction depends upon the passion, and especially of sensuality, a low and pagan way, unprofitable and fit only for the worldly-minded, and the habitual practice, on the other hand, of asceticism which is painful, unworthy and unprofitable. There is a middle path, O Bhikkhus, avoiding these two extremes, discovered by the Tathagat Buddha, a path which opens the eyes and bestows understanding, which leads to peace of mind, to the higher wisdom, to full enlightenment, to *Nirvana*."

And he explained to them the four truths concerning suffering, the cause of suffering, the destruction of suffering, and the way, which leads to the destruction of suffering. And the way was described to be eight folds. "This doctrine", Gautama said, "was not, O Bhikkhus, among the doctrines handed down. In Benaras in the hermitage of *Migdav*, the supreme wheel of the Empire of truth has been set rolling by the Blessed One-that wheel which not by any *Saman* or *Brahman*, not by any God, not by any *Brahma* or *Mar*, not by any one in the universe can ever be turned back."

The five former disciples of course were soon converted and were the first members of the order. Yasa, the son of the rich banker of Benares was his first lay disciple and the story of the conversion of this young man, nurtured in the lap of luxury and wealth, is worth repeating. He had three palaces, one of winter, one for summer, one for rainy season. One night he woke from sleep and found the female musicians still sleeping in the room with their dress and musical instruments in disorder. He became disgusted with what he saw and in moment of deep thoughtfulness said, "Alas, what distress, Alas ! What danger." And he left the house and went out. It was dawn and Gautama was walking up and down in open air and heard the perplexed and sorrowful young man exclaiming these worlds. He replied, "Here is no distress, Yasa. Here is no danger. Come here, Yasa. Sit down. I will teach you the truth." And Yasa heard the truth from the saintly teacher and became converted. Yasa's father, mother and wife all went to Gautama and listened to the holy truth. Yasa became a personal follower of Gautama, the other three remained his disciples.

Within 5 months of his arrival at Benares, Gautama had sixty followers. And now he called them together and dismissed them in different directions to preach the truth for the salvation of mankind. "Go you now, O Bhikkhus, and wander for the gain of the many, for the welfare of Gods and men. Let not two of you go the same way. Preach, O Bikkhus, the doctrine which is glorious in the end, in the spirit and in the letter; proclaim a consummate, perfect and pure life of holiness."

At Uruvela, Gautama converted three brothers named Kashyapa who worshipped fire in the Vedic form and had high reputation as hermits and philosophers. The eldest brother Uruvela Kashyapa and his pupils first flung their hair, their braids, their provisions and the things for *Agnihotr* sacrifice into the river and received the *Pavja* and *Upsanpada* ordination from the Blessed one. His brothers, who lived by the river Niranjara and at Gaya, soon followed the example. The conversions of the Kashyapa created a sensation and Gautama with his new disciples and 1000 followers walked towards Rajgrha, the capital of Magadha. News of the new prophet soon reached the king and *Sain·iya* Bimbisara surrounded by numbers of *Brahman* and *Vaeshya* went to visit Gautama. Seeing the distinguished Uruvela Kashyapa there, the king could not make out if that great Brahmin had converted Gautama or if Gautama had converted the Brahmin. Gautama understood king's perplexity and in order to enlighten him asked Kashyapa, "What knowledge have you gained, O inhabitant of Uruvela, that has induced you who were renowned for your penances, to forsake your sacred fire?" Kashyapa replied that he had seen the state of peace and took no more delight in sacrifices and offerings. The king was struck and pleased and with his numerous attendants declared himself an adherent of Gautama, and invited him to take his meal with him the next day.

The solitary wanderer accordingly went, an honoured guest, to the palace, of the king, and the entire population of the capital of *Magadha* turned out to see him. The king then assigned a *Bamboo* grove *Vain·uvan* close by for the residence of Gautama and his followers, and there, Gautama rested for some time. Shortly after Gautama obtained two renowned converts, Sariputta and Magellan.

The fame of Gautama had now traveled to his native town and his old father expressed a desire to see him once before he died. Gautama accordingly went to Kapilavastu, but according to custom remained in the grove outside the town. His father and relatives came to see him there. And the next day Gautama himself went into the town begging alms from the people who once adored him as their beloved prince and master. The story goes on to say that the king rebuked Gautama for this act, but Gautama replied it was the custom of his race. "But", retorted the king, "We are descended from an illustrious race of warriors and not one of them has ever begged his bread." "You and your family," answered Gautama, "may claim descent from kings; my descent is from the prophet, Buddha of the old."

The kings took his son to the palace where all the members of the family came to greet him except his wife. The deserted Yashodhara with a wife's grief and a wife's pride exclaimed, "If I am of any value in his eyes he will himself come; I can welcome him better here." Gautama understood this and went to her with only two disciples with him. And when Yashodhara saw her lord and prince enter a recluse with shaven head and yellow robes-her heart failed her; she flung herself to the ground, held his feet and burst into tears. Then remembering the impossible gulf between them, she rose and stood aside; she listened to his new doctrine and when subsequently Gautama was induced to establish an order of female mendicants Yashodhara became one of the first Buddhist nuns. Just at this time however she remained in her house but Rahula, Gautam's son, was converted. Gautam's father was aggrieved at this and asked Gautama to establish a rule that no one should in future be admitted to the order without his parents' consent. Gautama consented to this and made a rule accordingly.

On his way back to Rajgrha, Gautama stopped for some time at Anupiya, a town belonging to the Mallas. And while he was stopping there he made many converts both from the Kolian and from the Shakya tribe, some of whom deserve special mention. Aniruddh, the Shakya, went to his mother and asked to be allowed to go into the houseless state. His mother did not know how to stop him and so told him, "If Bhaddiya, the Shakya, will renounce the world thou also mayest go forth into the houseless state." Aniruddha accordingly went to Bhaddiya and it was decided that they should embrace the order in seven days. Chulvagga, the Buddhist Sutra says: So Bhaddiya, the Shakyaraj, and Aniruddha, and Ananda and Bhagu and Kimball and Devadatt just as they had so often previously gone out to the pleasure ground with fourfold array, and Upali the barber went with them, making seven in all. And when they had gone some distance, they sent their retinue back and crossed over to the neighboring district and took off their fine things and wrapped them in their robe and made a bundle of them and said to Upali the barber; "Do you now, Upali, turn back.

These things will be sufficient for you to live upon." But Upali was of a different mind so all the seven went to Gautama and became converts. And when Bhaddiya had retired into solitude, he exclaimed over and over, "O happiness! O happiness!" And on being asked the cause said: "Formerly, Lord when I was a king I had a guard, completely provided, both within and without my private apartments, both within and without and town and within the borders of my country. Yet though I was thus guarded and protected I was fearful.

anxious, distrustful and alarmed. But bow, Lord, even in the forest at the foot of a tree in solitude I am without fear or anxiety, trustful and not alarmed; I dwell at ease, subdued, secure, with mind as peaceful as an antelope."

Of these converts Ananda became the most intimate friend and companion of Gautama and after his death led the band of 500 monks in chanting the *Dharma* in the council of Rajagrha. Upali, though barber by birth, became an eminent member of the order and his name is often mentioned in connection with the *Vinyapitak*. Devadatta became subsequently the rival and opponent of Gautama and is even said to have advised Ajatashatru, the prince of Magadha, to kill his father Bimbisara and then attempted to kill Gautama himself.

After spending his second *Vas* or rainy season in Rajagrha Gautama repaired to *Sravasti*, the capital of Kosalas, where Prasenajit reigned as king. A wood called *Jaitvan* was presented to the Buddhists and Gautama often repaired and preached there. Gautam 's instructions were always delivered orally and preserved in the memory of the people like all the ancient books of India, although writing was known at this time.

The third *Vas* was also passed in Rajagrha and in the fourth year from the date of proclaiming is creed Gautama crossed the Ganges, went to Vaisali and stopped in the Mahavana grove. Thence he is said to have made a miraculous journey through the air to settle a dispute between the Shakyas and Kolians about the water of the Boundary River Rohini. In the following year he again repaired to Kapilvastu and was present at the death of his father, then 97 years old.

His widowed step-mother Prajapati Gautami and his no less widowed wife Yashodhara had now no ties to bind then to the world and insisted on joining the order established by Gautama. The sage had not yet admitted women to the order and was naturally most reluctant to do so. But his mother was inexorable and followed him to Vaisali and begged to be admitted. Ananda pleaded her case but Gautama still replied: "Enough, Ananda, Let it not please thee that women should be allowed to do so." But Ananda persisted and asked, "Are women, Lord, capable when they have gone forth form the household life and entered the homeless state under the doctrine and discipline proclaimed by the blessed one, are they capable of realizing the fruit of conversion, or of the second path or of *Arhat* ship?" There could be only one reply to this Honour to women has ever been a part of religion in India and salvation and heaven are not barred to the female sex by the Hindu religion. "They are capable," reluctantly replied Gautama.

And Prajapati and other ladies were admitted to the order as *Bhikkhunee* under some rules making them strictly subordinate to *Bhikkhu*. After this Gautama retired to Kausambi near Prayag (Allahabad).

In the sixth year after spending the rains at Kausambi Gautama returned to Rajagrha, and Kshema the queen of Bimbisara was admitted to the order. In same year Gautama is said to have performed miracles at Sravasti and went to heaven to teach *Dharma* to his mother who had died 7 days after his birth.

In the eleventh year Gautama converted the Brahmin Bharadvaja. In the next year he undertook the longest journey he had ever made and then preached the famous Mharahulsutan to his son Rahula, then 18 years old. Two years later Rahula was admitted in the order. In the fifteenth year he visited Kapilavastu again and addressed a discourse to his cousin Mahanama, who had succeeded Bhaddiya, the successor of Shuddhodana. Gautam's father-in-law Suprabuddha, the king of Koli, publicly abused Gautama for deserting Yashodhara but is said to have been swallowed up by the earth shortly after.

In the seventeenth year he delivered discourse on the death of Shrimati, a courtesan; in the next year he converted a weaver who had accidentally killed his daughter; in the following year he released a deer caught in a snare and converted the angry hunter who had wanted to shoot him; and in the twentieth year he similarly converted the famous robber Angulimala of the Chaliya forest.

For twenty-five years more Gautama wandered through the Gangetic valley, preached piety and holy life to the poor, the lowly and misguided, made converts among the high and the low, the rich and the poor and proclaimed his law wherever he went. He died at the age of 80. He lived 45 years from the date of his proclaiming the new religion.

3. This evening we proceed with the literature and philosophy of Buddhism:

(a) The form of Buddhism prevailing in Nepal, Tibet, China and Japan is called northern Buddhism, while the form prevailing in Ceylon and Burma is called Southern Buddhism. The Northern Buddhists furnished us with scanty material directly illustrating the religion in its earliest form in India. The sacred books of the Northern Buddhists are not included in any comprehensive common name and as far as is known none of then can be referred to the period immediately following on Gautam 's death. Kanishka, The king of Kashmir, convened a great. Kanishka, the king of Kashmir, convened a great council of the northern Buddhists in the first century after Christ, but the council instead of collecting together the sacred books of the Northern Buddhists wrote three great commentaries. The Lalitavistata, a most important work of the Northern Buddhists, is only a gorgeous poem; it is no more a biography of Gautama than the Paradise Lost a biography of Jesus. It was composed probably in Nepal in the second, third or fourth century, and the works on Buddhism which were then carried by Chinese pilgrims from India from century to century and translated into the Chinese language do not illustrate the earliest phase of Buddhism in India. And lastly, Tibet his drifted still further away from primitive Buddhism in India and has adopted forms and ceremonies, which were unknown to Gautama and his followers in the sixth century before Christ.

(b) On the other hand, the southern Buddhists furnish us with the most valuable materials. The sacred books of the Southern Buddhists are known by the inclusive name of the three *Pitakas* and there is evidence to show that these *Pitakas* now extant in Ceylon are substantially identical with the canon as settled in the council of Patna about 242 B.C.

(c) The three Pitakas are known as the *Suttee Pitakas*, the *Vinaya Pitakas* and the *Abhidhamma Pitakas*. The works comprised in the *suttee Pitakas* profess to record the sayings and doings of Gautama Buddha himself. Gautama himself is the actor and the speaker in the earliest worker of this *Pitakas* and his doctrines are conveyed in his own words. Occasionally one of his disciples is the instructor and there are short introductions to indicate where or when Gautama or his disciple spoke. But all through the *Suttee Pitakas Gautam's* doctrines and moral precepts are preserved professedly, in Gautam's own words.

The *Vinaya Pitakas* contains very minute rules for the conduct of monks and nuns who had embraced the holy order. Gautama respected the lay disciple *Upasak* but he held that to embrace the order was a quicker path to salvation. As the number of monks and nuns multiplied it was necessary to fix elaborate rules for their proper conduct and behaviour in the *Vihar* or monastery. As Gautama lived for nearly half a century after he had proclaimed his religion, there can be no doubt that he himself settled many of these rules.

The *Abhidhamma Pitakas* contains disquisition, on various subjects, like the conditions of life in different worlds, on the explanation of personal qualities, on the elements, the causes of existence etc. They have been miscalled metaphysics for early Buddhism knew little of metaphysics.

(d-e) Last time I said that the doctrine of four noble truths is the central point of Buddhist teaching. The substance of the teaching is, that life is suffering, the thirst for life and its pleasures is the cause of suffering the extinction or the thirst for life and its pleasure is the cause of suffering, the extinction of the thirst is the cessation of suffering, and that such extinction can be brought about by a holy life. We will discuss these four truths one after another.

(d) The first truth is the truth of suffering. As Gautama said: "Birth is suffering, decay is suffering, illness is suffering, death is suffering. Presence of objects we hate is suffering, not to obtain [objects] we desire is suffering. Briefly, the fivefold clinging to existence, *i.e.* clinging to the five aggregates, is suffering." What are those five aggregates? In Buddhist philosophy man is a compound of five aggregates. These are

Roop or the material aggregates-The first of the five. They include the four elements, earth, water, fire and air, five organs of sense, eye, ear, nose, tongue and body, five attributes of matter, form, sound, smell, taste and touch, two distinctions of sex, male and female, three essential condition, thought, vitality and space, two means of communication, gesture and speech, seven qualities of living bodies, buoyancy, elasticity, power of adaptation, power of aggregation, duration, decay chage. The second class of aggregates is *Vaidna* or sensations-the sensation of pleasure or pain. The third is *Sangya* or name. The fourth is *Sanskar* or the potentialities, which lead to good or bad results, and the fifth is *Vigyan* or knowledge. These five aggregates include all bodily and mental parts and powers of man and neither any one of them nor any group of them is permanent. It is repeatedly laid down in the *Pitakas* that none of these *skandhas* is soul. The body itself is constantly changing and so also each of the other aggregates. Man is never the same for two consecutive moments and there is within him no abiding principle whatever.

In *Sanyut Nikkei*, a Buddhist work, Buddha says: "mendicants, in whatever way the different teachers regard the soul, they think it is the five *skandhas* or one of the five. Thus mendicants, the unlearned, unconverted man who does not associate either with the converted or with the holy or understand their law or live according to it, such a man regards the soul either as identical with or as possessing or as containing or as residing in the material properties or sensations or in the other three *skandhas*. By regarding soul in one of these ways he gets the idea `I am'. Then there are the five organs of sense and mind and qualities and ignorance. From sensation produced by contact and ignorance the sensual, unlearned man derives the notions `I am' `This I exists', `I shall be' `I shall not be' etc. But now, mendicant, the learned disciple of the converted, having the same five organs of sense, has got rid of ignorance and acquired wisdom, and therefore the ideas `I am' etc. do not occur to him." This belief in self or soul is regarded in Buddhism so distinctly as a heresy those two well-known words in Buddhist terminology have been coined on purpose to stigmatize it. The first of these is *skayedithi*-the heresy of individuality-one of the three primary delicious which mush be abandoned at the very first stage of the Buddhist path of freedom. The other is *Atvad*, the doctrine of soul or self; it is classed with sensuality, heresy and belief in the efficacy of rites-as one of the four *upadans*, which are the immediate cause of birth, death, sorrow, lamentation, pain, grief and despair.

There is another Buddhist work called *Brahmjalsut,* in which Gautama discusses 62 different kinds of wrong belief; among them are those held by men who believe that the soul and the world are eternal, that there is no newly existing substance but these remain as a mountain peak unshaken, immovable, that living beings pass away, they transmigrate, they die and are born but these continue as being eternal. With regard to these Gautama says: "Upon what principle do these mendicants and Brahmins hold and doctrine of future existence? They teach that the soul is material or immaterial or is both or neither, that it is finite, or infinite or is both or neither, that it will have one or many modes or consciousness, that its perceptions will be few or boundless, that it will be in a state of joy, or of misery, or of both or of neither. These are the sixteen heresies teaching a conscious existence after death. Then there are eight heresies teaching that the soul material or immaterial or both or neither, finite or infinite or both or neither has an unconscious existence after death. And finally eight others which teach that the soul in the same eight ways exists after death in a state of being neither conscious nor unconscious." Lastly, he says: "Mendicants, that which binds the teacher to existence, *Tnha* or thirst, is cut off but his body still remains. While his body shall remain he will be seen by Gods and men, but after the termination of life, upon the dissolution of the body neither gods nor men will see him."

(e) So the first noble truth of Buddhism is that clinging to existence is misery. The second noble truth is the cause of misery. In Gautam's' words, "Thirst leads to rebirth accompanied by pleasure and lust-thirst for pleasure, thirst for existence, thirst for prosperity", And the third noble truth is, the cessation of suffering. It ceases with the complete cessation of thirst-a cessation, which consists in the absence of every passion-with the complete destruction of desire. The fourth truth is the noble truth of path, which leads to cessation of suffering. The holy eight-fold path is right belief, right meditation. The substance of the teaching is that without entering into any discussion into the origin and destiny of men one should lead a holy moral life and that will lead him to the summum bonum.

(f) On the eve of his death Gautama called together his brethren and appears to have recapitulated the entire system of morality under seven heads and these are known as the seven jewels of the Buddhist Law.

"Which then, O Brethren, are the truths which, when I had perceived, I made known to you, which, when you have mastered, it behooves you to practice meditate upon and spread, in order that pure religion may last long and be perpetuated, in order that it may continue to be for the good and happiness of the great multitudes, out of pity for the world, to the good and the gain and the weal of Gods and men? They are these:

1. The four earnest meditations.
2. The fourfold great struggle against sin.
3. The four roads to saint-ship.
4. The five moral powers.
5. The five organs of spiritual sense.
6. The seven kinds of wisdom.
7. The noble eight-fold path."

The four earnest meditations alluded to are: meditations on the body, the sensations, the ideas and the reason. The fourfold struggle against sin is the struggle to prevent sinfulness, the struggle to increase goodness. The fourfold struggle comprehends in fact a life-long, earnest, unceasing endeavour on the part of man towards more and more of goodness and virtue. The fourfold roads to saint-ship are the four means the will, the exertion, the preparation, and the investigation by which *Idhi* is acquired. In later Buddhism *Idhi* means occult powers but what Gautama meant was probably the influence and power which the mind, by long training and exercise, can acquire over body. The five moral powers and five organs of spiritual sense are faith, energy, thought, contemplation, investigation, joy, repose, and serenity. The eight-fold path we have referred to.

(g) It is by such prolonged self-culture, by the breaking of ten fetters doubt, sensuality etc. that one can at last obtain *Nirvana*.

Dhammapada says: "There is no suffering for him who has finished his journey and abandoned grief who has freed himself on all sides and thrown off all fetters. They depart with their thoughts well collected they are not happy within abode. Like swans those have left their lake, they leave their house and home. Tranquil is his thought, tranquil are his words and deeds, who has been freed by true knowledge, who has become a tranquil man."

It was generally believed that *Nirvana* meant final extinction and death, and Prof. Max Muller was the first to point out, which most scholars have now accepted [*viz.*] that *Nirvana* does not mean death but only the extinction of the sinful condition of mind, that thirst for life and its pleasure which brings on new births. *Nirvana* was not applied to any state after death, it was a term applied to a certain state of the life here. What Gautama meant by *Nirvana* is something attainable in this life, it is the sinless calm state of mind, the freedom from desires and passions, the perfect peace, goodness and wisdom which continuous self-culture can procure for man. As Rhys Davids puts it, "The Buddhist Heaven is not death and it is not on death but it is on a virtuous life here and now that the Pitakas lavish those terms of ecstatic description which they apply to Arhat-ship as the goal of the excellent way and to Nirvana as one aspect of it."

(h) But is there no future bliss, no future heaven beyond the virtuous life here and now for those who have attained *Nirvana*? This was a question, which often puzzled Buddhists and they often pressed their great master for a categorical answer. Gautama was an agnostic and to all questions about a future life after the attainment of *Nirvana* his reply was: "I do not know. It is not given me to know."

Malunkyaputta pressed this question on Gautama and desired to know definitively if the perfect Buddha did or did not live beyond the death. Gautama inquired, "Have I said `come Malunkyaputta and be my disciple, I shall teach thee whether the world is everlasting or not everlasting?" "That thou last not said, sire", replied Malunkyaputta. "Then", said Gautama, "do not press the inquiry."

Once king Prasenajit of Koshala during a journey between his two chief towns' Saketa and Shravasti met the nun *Khema* renowned for her wisdom. The king paid his respect to her and said, "*Venderable* lady, does and perfect one exist after death?" She replied, "The Exalted one, O great king, has not declared that the perfect one exists after death." "Then does the perfect on not exist after death, Venerable lady?" inquired the king. But Khema still replied, "This also, O great King, the Exalted one has not declared that the perfect one does not exist after death."

This shows that Gautam's religion was a perfect agnosticism, which did not and could not look beyond *Nirvana*. We know that according to Gautam's theory there is nothing permanent in man, that every particle mental, spiritual or physical, perishes every moment and new aggregates come into existence by reason of the influence left by the *karma* or action of the former aggregates. Everything is momentary, and if a man leads a perfectly holy life he would not collect new karma which will lead him into new birth; and therefore the aggregates of which he is composed come to an end without the new aggregates coming into existence. So although Gautama might not have said in so many words that the future state after *Nirvana* is a state of annihilation, still the natural conclusion is that the state must be that of total annihilation. In an article in the Lucifer of march 1874, Mr. G. R. Meads tries to save Buddhism from the charge of propounding a theory of annihilation and quotes a passage by Col. Olcott sanctioned by the High Priest of Ceylon. He says that although soul according to Buddhism is impermanent and changeable, still there is in man the permanent part called spirit. He says, "Buddhism does not deny the impressible nature of an ultimate spiritual reality in man, of a true transcendental subject, of an immortal changeless self." Now this self or transcendental subject has been know in all Indian philosophy by the name of *Atma*. With reference to Brahma Gautama has distinctly said in Tevijjia Sutta that the talk of the Brahmins about that Brahma is foolish talk and that there existed no such state as Brahma with reference to Brahma, I have already quoted Gautama as saying that it is heresy to say that there is any such thing as Brahma Soul and spirit; Atma and Brahama are all identical in Indian philosophies and an attempt to put into the mouth of Gautama views which he never maintained is fruitless attempt.

(i) If a man does not attain, while he is living, the state of *Nirvana* he is liable to future birth. Gautama did not believe in the existence of the soul, but nevertheless the theory of transmigration of souls was too deeply implanted in the Hindu mind to be eradicated and Gautama therefore adhered to the theory of transmigration without accepting the theory of soul! But if there is no soul, what is it that undergoes transmigration? The reply is given in the Buddhist doctrine of *karma*, which in its result corresponds to the Jaina and Hindu doctrines of *Karma* but in its foundation is entirely different from them. The doctrine is that *karma* or the doing of a man cannot die but must necessarily lead to its legitimate result. And when a sentient being dies a new being is produced according to the *karma* of the being that is dead. The cause which produces the new being is *Trishna* (thirst) or *upadan* (grasping). Sensation originates in the contact of the organs of sense with the exterior world; from sensations springs a desire to satisfy a felt want, a yearning, a thirst. From thirst results a grasping after objects to satisfy that desire, that grasping stage of mind causes a new being not, of course, a new soul, but a new set of *skandh*, a new body with mental tendencies and capabilities). The *karma* of the previous set of *skandh* or sentient being then determines the locality, nature and future of the new set of *skandh* or the new sentient being. Gautama said that in his philosophy four things are incomprehensible. The first is the effects of *karma*. And from what I have said just it is plain that the doctrine of *karma* as propounded by Gautama is an incomprehensible mystery.

(j) But the theory of transmigration was not the only theory, which Gautama accepted from the ancient religion and adopted, in a modified form into his own religion. The whole of the Hindu pantheon of the day was similarly accepted and similarly modified to suit his cardinal idea, the supreme efficacy of a holy life. The innumerable Gods of Rig-*Veda* were recognized but they were not supreme. Brahma, the supreme deity of the Upanishads, was recognized but was not supreme. Holy life alone was supreme and in preaching that doctrine. Buddha did an immense good; he raised goodness attainable by man above the gods and nature powers of Brahmins.

(k) How did Gautama deal with the caste system of the Brahmins? He respected a *Brahman* or *Shramana*, but he respected him for his virtue and learning, not for his caste which he altogether ignored. When two Brahmin youths, Vasishtha and Bharadwaja, began

(j) But the theory of transmigration was not the only theory, which Gautam accepted from the ancient religion and adopted, in a modified form into his own religion. The whole of the Hindu pantheon of the day was similarly accepted and similarly modified to suit his cardinal idea, the supreme efficacy of holy life. The innumerable gods of Rig-*Veda* were recognized but they were not supreme. Brahma, the supreme deity of the Upanishads, was recognized but was not supreme. Holy life alone was supreme and in preaching that doctrine Buddha did an immense good; he raised goodness attainable by man above the Gods and nature-powers of Brahmins.

(k) How did Gautama deal with the caste system of the Brahmins? He respected a Brahman, Sherman or Arhat, but he respected him for his virtue and learning, not for his caste, which he altogether ignored. When two Brahmin youths, Vasishtha and Bharadwaja, began to quarrel on the question "How does one become a Brahmin?" and came to Gautama for his virtue and learning, not for his caste which he altogether ignored. When two Brahmin youths, Vasishtha and Bharadwaja, began to quarrel on the question "How does one become a Brahmin?" and came to Gautama for his opinion Gautama delivered to them a discourse in which he emphatically ignored caste and held that a man's distinguishing mark was his work, not his birth.

(l) Gautama not only expressed his pronounced disapprobation against the Hindu case system he also exclaimed against the Vedic rites, which were practiced according to the injections of the ceremonial works. In place of such rites he enjoined a benevolent life and conquest of all passions and desires, and he recommended a retirement from the world as the most efficacious means for securing this end. The recommendation was followed and led to the Buddhist monastic system.

(m) And lastly, although Gautama himself disapproved of philosophical discussion, a system of Buddhist philosophy soon arose on the lines laid down by him; it ignored the existence of soul and maintained living creatures to be only assemblages of *skandhas* or aggregates; it knew of no state of future existence for those who attained *Nirvana*.

(n) What was it then that the Buddhists worshipped? What was the concrete form which Gautam's religion took in its early career before vast monasteries and an unwieldy priesthood replaced the primitive faith? What was the actual form of worship, which drew and engaged the multitude, which could not all have practiced or worshipped the abstract idea of a holy life? The reply is simple. For centuries, the people worshipped holiness and virtue as typified in the life of Gautama. They revered the memories of the great Teacher, they worshipped his invisible presence. The sculptures at Sanchi, at Amaravati, Barhut and other places represent homage paid to tree, to serpent, to the wheel or to the umbrella, but in every case the object represents the presence of Buddha.

(o) The moral precepts of Buddha are so well known that we shall pass over them and go at once to the history of Buddhism after Gautam's death. According to the Pali Scriptures, Buddha's death took place in 543 B.C. but the European scholars put it in 477 B.C. We are told in Chullavagga that on the death of Gautama, the venerable Maha Kashyapa proposed, "Let us chant the Dhamma and *Vinaya*" The proposal was accepted and 499 *Arhat* were selected for the purpose and

Ananda, the faithful friend and follower of Gautama, completed the number 500. And so they went up to Rajagrha to chant together the Dhamma and Vinaya. Upali, who was barber before, was questioned as the great authority on Vinaya and Ananda, the friend of Gautama, was questioned as the authority on Dhamma. This was the Council of Rajagrha held in the year of Gautam's death to settle the sacred text and fix it on the memory by chanting it together.

A century after the death of Gautama, a second council of 700 was held at Vaisali to settle disputes between the more and the less strict followers of Buddhism. It condemned a system of ten indulgences, which had grown up, but it led to the separation of the Buddhists into two hostile parties who afterwards split into 18 sects. During the next 200 years Buddhism spread over northern India. About 257 B.C. Ashoka, the king of Magadha, became a zealous convert to this faith. He founded many religious houses and his kingdom is called the land of monasteries.

Jainism

1. For this, the last lecture of the course, the subject that I have selected is Jainism, and I shall condense as much as possible the things that might be said on the subject.

Any philosophy or religion must be studied from all standpoints, and in order thoroughly, know what it says with regard to the origin of the universe, what its idea is with regard to God, with regard to the soul and its destiny, and what it regards as the laws of the soul's life. The answers to all these questions would collectively give us a true idea of the religion or philosophy. In our country religion is not different from philosophy, and religion and philosophy do not differ from science. We do not say that there is scientific religion or religious science; we say that the two are identical. We do not use the word religion because it implies a binding back and conveys and idea of dependence, the dependence of finite being upon an infinite, and [the idea that] in that dependence consists the happiness or bliss of the individual. With the Jainas the idea is a little different. With them bliss consists not in dependence but in independence; the dependence is in the life of the world and if that life of the world is a part of religion then we may express the idea by the English word, but the life which is the highest life, is that in which we are personally independent, so far as binding or disturbing influences are concerned. In the Highest State the soul, which is the highest entity, is independent.

2. This is the idea of our religion. The first important idea connected with it is the idea of universe. Is it eternal or non-eternal? Is it permanent or transitory? Or course, there are so many different opinions on the subject, but with these opinions I am not concerned in this lecture; I am only going to give the idea of the Jaina philosophy.

We say that we cannot study any idea unless we look upon it from all standpoints. We may express this idea by symbols or forms; we have expressed it by the story of the elephant and the seven blind men who wanted to know what kind of animal the elephant was, and each, touching a different part of the animal, understood its form in so many different ways and thereupon became dogmatic. If you wish to understand what kind of animal an elephant is, you must look upon it from all sides, and so it is with truth. Therefore we say that the universe from one standpoint is eternal and from another one-eternal. The totality of the universe taken as whole is eternal. It is a collection of many things.

That collection contains the same particles every moment, therefore as collection it is eternal; but there are so many parts of that collection and so many entities in it, all of which have their different states which occur at different times and each part does not retain the same state at all times. There is change, there is destruction of any particular form, and a new form comes into existence; and therefore if we look upon the universe from this standpoint it is non-eternal. With this philosophy there is no idea, and no place for the idea, of creation out of nothing. That idea, really speaking, is not entertained by any right-thinking people. Even those who believe in creation believe from a different standpoint than this. It cannot come into existence out of nothing, but is an emanation coming out of something. The state only is created. This book in a sense is created because all its particles are put together, having been in a different state. The form of the book is created. There was a beginning of this book and there will be an end. In the same manner, with any form of matter, whether this form lasts for moments or for centuries, if there was beginning of this book and there will be an end. In the same manner, with any form of matter, whether this form lasts for moments or for centuries, if there was a beginning there must be an end.

We say there are both preservation and destruction in the many forces working around us all these forces are working ever moment in the midst of us and around us, and the collection of these entities is called by the Jainas 'God'. The Brahman Represent it by the syllable OM; the first sound in this word represents the idea of creation, the second of preservation and the third of destruction. All these are energies of the universe and taken as a whole they are subject to certain fixed laws. If the law are fixed why do people bow down to these energies? Why do they consider the collective energy as a god or as God? There is always an idea of the power to do evil in the beginning of this conception.

When railroads were first introduced into India ignorant people who did not know what they were, who had never seen in their lives that a car or carriage could be moved without the horse or the ox, thought that there was some divinity in the engine, some God or Goddess and some of them even bow down before the car; and even to this day you will find in some parts of India, among the pariahs or low class, that there are people who entertain this idea. So to these energies in our primitive state we are liable to attribute personality; and after a long course of development we symbolize our thought in the form of pictures and explain them in that way to make them more intelligible to others.

In the ancient times there was not rain but a rainer, not thunder but a thunderer, and in that way, personality is attributed, or living consciousness and character, to those forces. There may be conscious entities in these forces, as there may be living entities on the planets, but these forces themselves are not living entities. This, however, expresses the idea in the beginning; these energies were classed as creative, preservative and destructive, and these three entities were considered to be component parts of one entity called Brahma by the Hindus. Really, creation in this is in the sense of emanation, preservation is used in the sense of preserving the form, and destruction in the sense of destroying the form.

The idea of matter is something that can be handled or perceived by the senses and the energies must be material energies, as cohesion, magnetism, electricity, but to consider these God would be the most materialistic idea, and therefore the Jainas discard this idea so far as the Godhead or Godlike character is concerned. They of course admit the existence of these energies, that they are indeed to be found everywhere, but they are subject to fixed laws which cannot be interfered with by any person, not that these energies consciously influence our destinies with regard to good and evil. To say that they do so influence us is only to show our ignorance with regard to their laws.

These energies collectively we call substantiality. There are innumerable qualities and attributes in matter itself, and they manifest themselves at different times and ways. We are not able selves at different times and ways. We are not able without further development to know what energies are inherent in matter, and when any new thing comes to view we are surprised, and whatever is surprising is considered to be something coming from divinity; but where we understand scientific principles the surprise is removed and it is all as simple as the daily rising and setting of the sun. Thousands of years ago the different phenomena of nature were considered in different parts of the world to be the working of different Gods and Goddesses, but when we understand science these phenomena become simple and the idea of theses beings as characters of the highest spiritual power goes away.

3. 'What is the God of the Jainas?' you will ask. I have only told you what he is not. I will now tell you what it is. We know that there is something besides matter; we know that the body exhibits many qualities and powers not to be found in

ordinary material substances, and that the some thing which causes this, departs from the body at death. We do not know where it goes; we know that when it lives in the body the powers of the body are different from what they are when it is not there. The powers of nature can be assimilated to the body at death. We do not know where it goes; we know that when it lives in the body the powers of the body are different from what they are when it is not there. The powers of nature can be assimilated to the body when that some thing is there. That entity is considered by us the highest and it is the same inherently in all living beings. This principle common to us all is called divinity. It is not fully developed in any of us, as it was in the saviors of the world, and therefore we can them divine beings. So the collective idea derived from observations of the divine character inherent in all beings is by us called God. While there are so many energies in the material world and in the spiritual and putting those two energies together we give them the name of nature we separate the material energies and put them together; but the spiritual energies we put together and call them collectively God. We make a distinction and worship only the spiritual energies. Why should we do so? A Jaina verse says, "I bow down to that spiritual power or energy which is the cause of leading us to the path of salvation, which is supreme, which is omniscient; I bow down to the power because I with to become like that power." So where the form of the Jaina prayer is given the object is not a receive anything from that entity or from that spiritual nature, but to become like that power." So where the form of the Jaina prayer is given the object is not to receive any thing from that entity or from that spiritual nature, but to become one like that; not that spiritual entity will make us, by a magic power, become like itself, but by following out the ideal which is before our eyes, we shall be able to change our own personality; it will be regretted, as it were, and will be changed into a being which will have the same character s the divinity which is our idea of God. So we worship God, not as being who is going to give up something, not because it is going to do something or please us, not because it is profitable in way; there is not any idea of selfishness; it is like practicing virtue for the sake of virtue and without any other motive.

4. (a) Now we come to the idea of soul. The ordinary idea of soul substance is that in order for thing to exist it must have formed, it must be perceived by the senses. This is our ordinary experience. Really speaking it is the experience only of the sensuous part of the being, the lowest part of the human entity, and from and experience we derive conclusions and think that these conclusions apply to all substance. There are substances, which cannot be perceived by the senses; there are subtler substances and entities and these can be known only by the consciousness, by the soul. Such a substance which cannot be seen, heard, tasted, smelt or touched, is a substance which need not occupy space and need not have any tangibility, but, it may exist although it may not have nay form. Sight is an impression made on the nerves of the eye by vibrations sent forth form the object perceived and this impression which we call sight, if there are no vibrations coming out of the object, is of course not produced; but if this substance influences us in certain ways the implication is that there is something moving or producing vibrations, and these cannot exist unless there is some material substance which is vibrating. The very fact that something is moving in some way and influences us in some peculiar way implies that there is something material about this. If there are no vibrations the substance is not material. It need not exist in a form, which will give us the impression of any colour, smell, etc. There is nothing, which can partake both the attributes of soul and of matter; the attributes of matter are directly contrary to those of the soul. While one has its life in the other, it does not become the other.

How can that soul live in matter when its attributes are of a different nature? By our own experience we know that, we are obliged to live in surrounding which are not congenial to us, Which are not of our own nature. People feel that they are not related to their surroundings, there must be some reason for their being obliged to live in those surroundings, but there must be a reason in the intelligence itself; it cannot be in the material substance. We know that this is fact, because intelligence cannot proceed from any thing, which is purely material. No material substance has given any evidence of having possessed intelligence; it might have done so when there was life in it, but without this it has no intelligence. That intelligence is, we are quite sure, influenced by material things, but it does not arise from the material things. Persons of sound intelligence take a large dose of some intoxicating drink and the intelligence will not work at all. Why should this material thing influence the immaterial, the soul? The soul thinks that the body is itself and therefore anything, which is done to the material self, is supposed by the real self to be done to it. That is where the Christian scientists and the Jaina philosophy will agree; that if the soul thinks that the body is real self anything done to the body will be considered by the soul to be done to the soul, and therefore what happens to the body will be felt by the soul; but if the soul for a moment thinks that the body is not the self but altogether different and a stranger to the soul, for that reason no feeling of pain will exist; our attention is taken away in some other direction shows that the self is something higher than the body. Still under ordinary circumstances the soul is influenced by the body, and therefore we are to study the laws of the body and soul so as to rise above these little things and proceed on our path to salvation or liberation, which is the real aspiration of the soul. There is power of matter itself, but that power is lower than the power of the soul. If there was no power at all in the body or in matter, the soul would never be influenced by it, for mere non-existence will never influence anything; but because there is such a thing as matter when the soul thinks that that the soul would never be influenced by it, for mere non-existence will never influence anything; but because there is such a thing as matter, when the soul thinks that there is a power of the body and a power of the matter, these powers will influence it. Bodily power as we see it is on account of the presence of the soul. There is a power in matter, as cohesion etc., and this will work although the soul does not think anything about it. If the moon revolves around the earth there are some forces inherent in the earth and moon. What I mean to say is that the influence of these material powers on soul powers depends on the soul's readiness or willingness to submit to these powers. If the soul takes the view that it will not be influenced by any thing, it cannot be so influences.

(b) This being the soul's nature, what is its origin? Everything can be looked upon from two standpoints, the substance and the manifestation. If the state of the soul itself is to be taken into consideration that state has its beginning and its end. The state of the soul as living in the human body had a beginning at birth and will have an end at death, but it is a beginning and an end of the state, not of the thing itself. The soul taken as a substance is eternal; taken as a state every state has its beginning and end. So this beginning of a state implies that before this beginning there was another state of the soul. Nothing can exist unless it exists in some state. The state may not be permanent, but the thing must have a state at all times. if therefore the present state of the soul had beginning, it had another state before the beginning of this state, and after the end of this state it will have another state. So the future state is something that comes out of or is the result of the present state. As the future is to the present, so the present is to the past. The present is only the future of the past. What is true with regard to the future state is true with regard to the past and present state. The acts of the past have determined our present state, and if this is true the acts of the present state must determine the future state.

This brings us to the doctrine of rebirth, transmigration of souls, metempsychosis, reincarnation, etc. as they are variously known. First take incarnation, which means literally becoming flesh; and, really that which is spirit is always spirit or soul. The spirit does not become flesh. If reincarnation means to become flesh there can be no reincarnation, but if it means simply the life in flesh for a short time, then there is reincarnation. Reincarnation means also to be born in some state again and again. Metempsychosis means in Greek only change; that the animal itself, body and soul, everything together, is changed into some other thing and so on. That is the idea of metempsychosis. Transmigration of souls is, especially in the idea of the Christians, the idea of the human soul going into the animal body, as if this were a necessity. But that is not the real idea; the real idea is simply going from one place to another or from one body to another, but not necessarily going from the human body to the animal body, but simply travelling. It implies the idea of form. Nothing can travel unless it has form and occupies space and is material; so in our philosophy we reject all these terms if that is the idea connected with these terms, and use the idea of rebirth; that is, the soul is born in some other body, and the birth does not imply the same conditions [as those] applying to the human birth.

There are certain conditions in which human beings are born; the seed itself takes several months to ripen and then there is the birth. This may be due to certain acts or forces, which are generated by human beings? These are in a condition to be observed by beings whose forces will take them to some other planet, and we say that there is another condition of birth there. There is no necessity for gestation and fecundation. The karmic body has in itself many powers and has a force to take to itself another body, which is in the case of the human beings a gross body, but in the case of other beings a subtle body is generated and this body is changeable so far as its form and dimensions are concerned. Therefore, if the forces generated, while we live any kind of life are of different kinds then in the case of some being it may be necessary that he should be born in the human condition and pass through the actual conditions which must be obeyed if the human being is to be born, while if the forces generated are different in their character he may be born on some other planet, where birth is manifested in different way, without any necessity of the combination of the male and the female principle.

There are so many different planes of life that the mere study of the human life ought not to be made to apply to all the affairs of life. We have studied only a few forms of the life of animals, human beings, etc., but that is only the part which under the present development of our science of our eyesight even, we are able to study. We are not able to study other forms of life, innumerable in the universe, and therefore we ought not to apply the laws thus discovered to all forms of life.

Our study is introspective because our idea is that soul is able to know everything under the right circumstances. The knowledge acquired under these conditions is of the sounder nature and a more correct kind because the obstacles, which come in the way science, are not there. Science has to commit mistakes and think that they do not; still knowledge is derived from inferences which we draw from certain premises which may not be right or if the premises are right the inference may be wrong. We do not mean to say that there are always mistakes in the knowledge, which is acquired through sensation or through matter, but sometimes it is possible, and while it may be correct knowledge in many cases we cannot rely on that.

The highest knowledge is immediate knowledge, derived by the soul without the assistance of any external thing, and the knowledge of liberated souls, and also the knowledge of human beings who are just on the point of being liberated, or have passed through the course of discipline, mental, moral and spiritual, and have nearly exhausted past forces, at the same time generating spiritual forces, and on account of discipline and sees everything when this state is arrived at; it knows means that it is something, some reality, and there can be no reality unless it can distinguish itself from other realities. Only the one universal thing could not know itself, because knowledge implies comparing one with another, and if that itself, because knowledge implies comparing one with another, and if that is not done there is no individuality.

We say therefore that the soul in its highest existence knows, that it is perfectly separate form other things, so far as experience and knowledge are concerned, but in so far as its nature is concerned, so long as there is a sense of separateness, there is no occasion or opportunity for the soul to rise higher because when the soul thinks that it is living a different existence for its own sake it is considering its own self to be different from another person and thinks that this is its own and a part of its nature, its own being, and therefore anything done in regard to these surroundings will benefit or injure its own. It even thinks that its very life consists in doing goods and in loving other souls and taking active measures for carrying into effect the very plan of the sole (Those souls?). Then it comes higher, and ultimately reaches the highest condition. The condition of the soul, as I have said, is the highest in which there is perfect consciousness, there is infinite knowledge and infinite bliss; we express these three ideas in Sanskrit as existence infinite, bliss infinite and knowledge infinite. That condition of the soul cannot be described by us because description is something which proceeds from a finite mind and when the soul becomes infinite no finite mind and when the soul becomes infinite no finite mind can fully express the conditions of that infinite state. The attributes we give therefore to that condition of the soul are always full of comprehension. We shall always leave out many things; we have not the power to express all our thoughts. How can we express, then, this state of a soul, which so far as its power and knowledge are concerned is infinite?

The Jainas have studied the nature of the soul and the universe from these standpoints and have derived a beautiful principle, and so far as this is concerned there is this difference between this country and other countries and other religions, they can understand all these from these standpoints. The Bible says, 'Thou shalt not kill', and Jainas practice universal love so that this also means that we should not kill any beings. If we say that the Bible does not mean that we take away a part of the bible. Why should we interpret the laws of any religion from the narrowest standpoint? We should take into consideration the nature, attributes and working of all things. We cannot derive laws, which are to be applied to the whole universe simply by our observation of a part of the conscious nature of the universe.

If you wish to state correctly the nature of the universe you will study the nature of all the different parts of the universe and then the laws will be applicable to all parts of it. We think that we are superior to other things because our tenants who live on the ground floor are inferior to us, but we have no right therefore to crush those tenants, who later on will acquire the right to inhabit the second and third floors and finally the highest floor. One living on the highest plane has no right to crush those who live on the lowest plane. If one thinks that he has a right to do this, which he has, not sufficient strength to live without destroying life, our philosophy says that it is still sin to destroy life, and it remains only to choose the lowest form, the less evil, We will in business take such a kind of business as will yield the most profit and will cause us to lose the least, in which we have the less liabilities; and the highest condition will be that in which we have no liabilities and no creditors, the state in which we may live without any creditors or in a perfectly free condition. That is the liberated condition.

5. The idea of *Karma* is very complicated. I have told you something so it in my former lectures. The one chief point is that that theory is not the theory of fatalism, not a theory in which the human being is tied down to some, one, bound down by the force of something outside itself. In one sense only will there be fatalism; if we are free to do many things we are also not free to do other things, and we cannot be freed from and results of our acts. Some results may be manifested in great strength, others very weakly; some may take a very long time and others a very short time; some are of such a nature that they take a long time to work out, while the influence of others may be removed by simply washing with water and that will be the case in the matter of acts done incidentally without any settled purpose or any fixed desire. In such a case with reference to many acts we may counteract their effects by willing to do so. So the theory of *karma* is not in any sense a theory of fatalism, but we say that all of us are not going to one goal without any desire on our part, not that we are to reach that state without any effort on our part, but that our present condition is the effect of our acts, thoughts and words in the past state. To say that all will reach the prefect state merely because some one has died that they might be saved, merely from a belief in this person, would be a theory of fatalism, because those who have lived a pure and virtuous state and have not accepted a certain theory will not reach the perfected state simply for that reason and no other the faith in saviors is simply this, that by following out the divine principle which is in our selves when this is fully developed we also shall

become Christ's, by the crucifixion of the lower nature on the altar of the higher. We also use the cross as a symbol. All living beings have to pass through or evolve from the lowest, the monadic, condition to the highest state of existence and cannot reach this unless they obtain possession of the three things necessary; right belief, right knowledge and right conduct. The right, belief, really speaking, is not that there is no passing through forms after death, but the soul keeps progressing always in its own nature without any backward direction at all. We have expressed this in clear language without any parables or metaphors, but when we preach these truths to the ignorant masses some story or picture might be necessary for them and after that the explanation of the real meaning; as well have all allegory in the Pilgrim's Progress. It is just like reaching the Celestial City in that book, but we must all understand that these things are parables. Others may need music to assist their religion, but when we understand the esoteric meaning which underlies all religion there will be no quarrelling and no need of names or of forms; and this is really the object of all religions.

Bibliography

Agrawala, V. S.: *Shiva Mahadeva: The Great God*, Veda Academy, Varanasi, 1966.

Alladi Mahadeva Sastri: *The Bhagavad Gita: with the Commentary of Sri Sankaracharya*, Samata Books, Madras, 1977.

Apte, V. M.: *Social and Religious Life in the Grihya Sutras*, Rawat Press, Bombay, 1954.

Aurobindo, Sri: *Vyasa and Valmiki*, Acharya Press, Pondicherry, 1956.

Avalon, Arthur: *Hymn to Kali (Karpuradi-Stotra)*, Ganesh and Co., Madras, 1965.

Baillie, John: *The Belief in Progress*, Oxford University Press, London, 1960.

Baird, Robert: *Religion in Modern India*, Manohar, New Delhi, 1981.

Basham, A.L.: *A Cultural History of India*, Clarendon Press, New York, 1975.

Bernard, Theos: *Hatha Yoga*, Rider and Co., London, 1969.

Bernasconi, Robert : *Heidegger in Question : The Art of Existing*, Humanitas Press, New Jersey, 1993.

Berreman, Gerald D.: *Hindus of the Himalayas*, University of California Press, Berkeley, 1963.

Bharati, Agehananda: *The Ochre Robe*, University of Washington Press, Seattle, 1962.

Bhattacharyya, Vidhusherhar : *The Agamasaastra of Gaudapaada*, Calcutta University Press, Calcutta, 1943.

Caputo, John D. : *Demythologizing Heidegger*, Indiana University Press, Bloomington, 1993.

Cousineau, Robert H. : *Humanism and Ethics: An Introduction to Heidegger's 'Letter on Humanism', with a Critical Bibliography*, Beatrice Nauwelaerts, Paris, 1972.

Danner, Helmut : *Das Goettliche undder Gott bei Heidegger*, Anton Hain, Meisenheim, 1971.

Delp, Alfred : *Tragische Existenz: Aur Philosophie Martin Heideggers*, Herder, Freiburg, 1935.

Edwards, Paul : *Heidegger on Death: A Critical Examination*, The Hegeler Institute, Illinois, 1979.

Elliston, F. : *Heidegger's Existential Analytic*, Mouton Publishers, The Hague, 978.

Frings, Manfred S. : *Heidegger and the Quest for Truth*, Quadrangle Books, Chicago, 1968.

Fynsk, Christopher : *Heidegger : Thought and Historicity*, Cornell University Press, Ithaca, New York, 1986.

Grene, Majorie : *Martin Heidegger*, Hillary House, New York, 1957.

Haeffner, Gerd : *Heideggers Begriff der Metaphysik*, Berchmanskolleg, Muenchen, 1974.

Halliburton, David : *Poetic Thinking: An Approach to Heidegger*, University of Chicago Press, Chicago, 1981.

Hodge, Joannaa : *Heidegger and Ethics*, Routledge, London, 1995.

Ihde, Don : *Hermeneutic Philosophy*, Northwestern University Press, Evanston, 1971.

Iyer, K.A. Krishnaswami : *Vendaanta; The Science of Reality*, Ganesh and Co., Madras, 1930.

Jaeger, Alfred : *Gott nochmals Martin Heideggers*, Mohr, Tuebingen, 1978.

Kearney, Richard : *Modern Movements in European Philosophy*, Manchester University Press, Great Britain, 1986.

King, Magda : *Heidegger's Philosophy: A Guide to his Basic Thought*, Basil Blackwell, Oxford, 1969.

Kovack, George : *The Question of God in Heidegger's Phenomenology*, NUP, Evanston, 1990.

Krell, D.F. : *Intimations of Mortality: Time, Truth and Finitude in Heidegger's Thinking of Being*, Pennsylvania State University, University Park, 1986.

Levinas, Emmanuel : *Totality and Infinity: An Essay on Exteriority*, Martinus Nijhoff, The Hague, 1979.

Luegenbiehl, Heinz C. : *The Essence of Man: An Approach to the Philosophy of Martin Heidegger*, Microfilms, Ann Arbor, 1979.

Luijpen, W.A. : *Phenomenology and Humanism: A Primer in Existential Phenomenology*, Duquesne University Press, Pittsburgh, 1966.

Macaun, Christopher : *Martin Heidegger: Critical Assessments*, Routledge, London, 1992.

Manikath, Joseph : *From Anxiety to Releasement: In Martin Heidegger's Thought*, A.T.C., Bangalore, 1978.

Marx, Werner : *Heidegger and the Tradition,* Northwestern University Press, Evanston, 1971.

McDaniel, Jay B. : *Earth, Sky, Gods and Mortals: Developing an Ecological Spirituality*, Twenty-third Publications, Connecticut, 1990.

Mueller, F. Max : *The Six Systems of Indian Philosophy*, Logmans, Green and Co., London, 1912.

Organ, Troy Wilson : *The Self in Indian Philosophy*, Mounton & Co., London, 1964.

Perotti, J.L. : *Heidegger on the Divine: The Thinker, the Poet and God*, Ohio University Press, Athens, 1974.

Poeggeler, Otto : *Der Denkweg Martin Heideggers*, Neske. Pfullingen, 1983.

Radhakrishnan, S. : *Indian Philosophy*, George Allen & Unwin Ltd., London, 1941.

Rukavina, T.F. : *Heidegger as a Critic of Western Thinking*, University Microfilms, Ann Arbor, 1977.

Schmitt, Richard : *Martin Heidegger: On Being Human: An Introduction to 'Sein und Zeit'*, Random House Inc., New York, 1976.

Schrag, Calvin O. : *Existence and Freedom: Toward an Ontology of Human Finitude*, Northwestern University Press, Evanston, 1961.

Sinari, Ramkant A. : *The Structure of Indian Thought*, Charles C. Thomas Publisher, Illinois, 1970.

Taminiaux, Jacques : *Heidegger and the Project of Fundamental Ontology*, State University of New york Press, Albany, 1991.

Ussher, Arland : *Journey Through Dread*, The Delvin Adair Company, New York, 1955.

Vail, L.M. : *Heidegger and the Ontological Difference*, State University Press, Pennsylvania, 1972.

Versenyi, Laszo : *Heidegger, Being and Truth*, Yale University Press, New Haven, 1965.

Wisser, Richard : *Martin Heidegger in Gespraech*, Karl Aber, Freiburg, 1969.

Wood, David : *Heidegger and Language: A Collection of Original Papers*, Parousia Press, Coventry, 1981.

Zimmermann, Michael E. : *Eclipse of the Self: The Development of Heidegger's Concept of Authenticity*, Ohio University Press, London, 1981. allerary

Index

G

H

I

J

K

L

M

N

O

P

R

S

T

□□□